Cornelius Thiele

STM Characterization of Phenylene-Ethynylene Oligomers on Au(111) and their Integration into Carbon Nanotube Nanogaps

Experimental Condensed Matter Physics
Band 12

Herausgeber
Physikalisches Institut

Prof. Dr. Hilbert von Löhneysen
Prof. Dr. Alexey Ustinov
Prof. Dr. Georg Weiß
Prof. Dr. Wulf Wulfhekel

Eine Übersicht über alle bisher in dieser Schriftenreihe erschienenen Bände finden Sie am Ende des Buchs.

STM Characterization of Phenylene-Ethynylene Oligomers on Au(111) and their Integration into Carbon Nanotube Nanogaps

von
Cornelius Thiele

Dissertation, Karlsruher Institut für Technologie (KIT)
Fakultät für Physik
Tag der mündlichen Prüfung: 16. Mai 2014
Referenten: Prof. H. v. Löhneysen, Prof. W. Wulfhekel

Impressum

Karlsruher Institut für Technologie (KIT)
KIT Scientific Publishing
Straße am Forum 2
D-76131 Karlsruhe

www.ksp.kit.edu

Print on Demand 2014

ISSN 2191-9925
ISBN 978-3-7315-0235-7
DOI 10.5445/KSP/1000041811

I may not have gone where I intended to go,
but I think I have ended up where I needed to be.

Douglas Adams
The Long Dark Tea-Time of the Soul

Contents

1 Introduction **1**

2 Materials and Methods **3**

2.1 Oligo(phenylene ethynylene)s . . . 3
 2.1.1 OPE-A and OPE-B . . . 5
2.2 Scanning Tunneling Microscopy . . . 9
 2.2.1 Atomic Resolution Imaging . . . 12
 2.2.2 The Au(111) Surface . . . 12
2.3 Density Functional Theory . . . 14
2.4 Carbon Nanotubes . . . 17
 2.4.1 Synthesis of Carbon Nanotubes . . . 19
 2.4.2 Preparation of Nanotube Suspensions . . . 20
 2.4.3 Dielectrophoresis of CNTs . . . 21
 2.4.4 Electron-Beam Lithography . . . 22
2.5 Electron-Beam-Induced Oxidation . . . 24
 2.5.1 The Electron Microscope . . . 25
 2.5.2 Electron-Beam-Induced Oxidation of Carbon . . . 27
2.6 Helium-Ion-Beam Lithography . . . 29
 2.6.1 Ion Beam - Sample Interaction . . . 29
 2.6.2 Ion-Beam Lithography . . . 31

3 STM Study of OPE-A and OPE-B **33**

3.1 Surface Preparation . . . 33
 3.1.1 Dichloromethane on Au(111) . . . 34
3.2 Self-Assembly of OPE-A and OPE-B . . . 35
3.3 Single OPE Molecules . . . 37
 3.3.1 Moving Molecules on the Surface . . . 39
 3.3.2 Contacting with the STM tip . . . 41
3.4 Scanning Tunneling Spectroscopy . . . 46
 3.4.1 Conductance Maps . . . 50
3.5 Atomic Resolution Imaging . . . 54
3.6 Conformational Switching of Alkyl Chains . . . 58
3.7 DFT Calculations of OPE-A and OPE-B . . . 65
 3.7.1 Calculations of Molecules in the Gas Phase . . . 65
 3.7.2 Comparison with STM Measurements . . . 68
 3.7.3 Calculations including the Au(111) Surface . . . 74

4 Carbon Nanotube Nanogap Electrodes **79**
4.1 Nanogaps by Electron-Beam-Induced Oxidation 79
4.1.1 Sample Preparation 79
4.1.2 Experimental Setup 82
4.1.3 Electrical Characterization and Conditioning 82
4.1.4 Results of Nanogap Fabrication 84
4.1.5 Discussion and Conclusion 87
4.2 Nanogaps by Helium-Ion-Beam Lithography 90
4.2.1 Device Fabrication 90
4.2.2 Sputtering of Nanogaps 93
4.2.3 Characterization of Nanogaps 93
4.2.4 Estimation of Cross-Section 100

5 Direct Contacting of OPE-A **103**
5.1 Deposition of OPE-A into Nanogaps 103
5.2 Electrical Characterization 105

6 Conclusion and Outlook **107**

Bibliography **111**

1 Introduction

The field of molecular electronics strives for a better understanding of the mechanism of electronic transport in molecules, in order to shrink down the individual components of an integrated circuit to the size of a single molecule. To replace all parts of integrated circuits, the complex three-dimensional interconnects between their functional units need to be addressed as well. Initially made from aluminum, the industry has switched to copper interconnects [4] because of their lower specific resistance. For electronic circuits on a molecular scale, these bulk metal interconnects need to be replaced by molecular wires. The term *molecular wire* thereby denotes a discrete organic molecule, resembling a wire due to its rod-like structure and because of its intended purpose of carrying an electric current. In the literature, several variants of conjugated aromatic molecules are discussed as molecular wires [5].

The first experiment to measure a current through a single layer of organic molecules only was performed on fatty acid monolayers [6]. Later it was discovered that mechanically controlled break junctions [7] can be used to contact molecules. Through thousands of cycles of breaking the contact and re-establishing it, conductance histograms of a selected molecule can be measured [8]. Because the junction geometry and the coupling to the molecule change with each cycle, only such a statistical analysis is possible.

The scanning tunneling microscope (STM), invented by Binnig and Rohrer at IBM Research Zurich in 1979 [9, 10], is based on vacuum tunneling of electrons between a sharp tip and a surface. It allows detailed imaging and furthermore the analysis of the electronic properties of surfaces and the molecules on top of them. Advances in the field have led to the development of instruments operating at cryogenic temperatures in ultra-high vacuum (UHV), allowing detailed views of single molecules in a virtually contamination-free environment.

This work sets out to span the bridge between a detailed STM analysis of an organic molecular wire and direct electrical measurements on it by placing it into a device geometry. To that end, a liquid-helium cooled UHV STM was used to study two variants of oligo(phenylene ethynylene) (OPE) molecules, molecular wires up to $\approx 5\,\mathrm{nm}$ in length, on a Au(111) single-crystal surface. Self-assembled structures of both molecules as well as single molecules were analyzed at sub-molecular resolution. Direct images of the chemical structure of the molecules could be obtained at bias voltages in the range of the HOMO-LUMO energy gap. Spectroscopic images of the molecular orbitals and I-V spectra were recorded and compared to density functional theory calculations. Furthermore, a conformational switching behavior of the side groups of both molecules was discovered and analyzed in detail. Results of the STM study are presented in Chapter 3.

Our efforts towards directly contacting the molecules by carbon nanotube nanogap electrodes is presented in Chapter 4. Carbon nanotubes can be thought of as rolled-up sheets of sp^2-hybridized carbon, with their electronic properties depending on their roll-up vector. Recently, sorting methods were devised to separate metallic from semiconducting and eventually even single chiralities of carbon nanotubes from each other [11]. Because of their one-dimensionality and their excellent electronic properties as compared to a thin strip of metal of the same dimensions, metallic carbon nanotubes are ideal electrodes for experiments on single molecules [12]. Covalently bridged contacts [13] as well as π-orbital-mediated bond geometries have been reported to successfully contact single molecules [14].

However, the fabrication of these electrodes typically involves either current-induced oxidation or lithographic cutting with an oxygen plasma. Both methods are difficult to implement and do not have a high yield of sufficiently small gaps [14, 15]. In the course of this work, two methods for the fabrication of nanogaps in carbon nanotubes were explored, with the goal of contacting a single OPE molecule by inserting it into a suitably-sized nanogap.

The first method is electron-beam-induced oxidation inside a scanning electron microscope, using the electron beam of the instrument to induce oxidation at the focal point. This technique is covered in Sec. 4.1. Nanogaps with an average size of 20 nm were produced, which were still too large as electrodes for a single molecule. However, helium-ion microscopes recently became available, which image surfaces by scanning a focused helium-ion beam across it. These ions sputter material as well, and through helium-ion sputtering it was possible to manufacture nanogaps with a size of only 2.8 nm into single carbon nanotubes. This method is presented in Sec. 4.2. The carbon nanotubes were embedded in a device geometry, allowing first electrical measurements on single OPE molecules deposited into these nanogaps, which are demonstrated and discussed in Chapter 5.

To first acquaint the reader with the historical and theoretical background of the materials and methods used in this work, they are introduced in the following chapter. This includes oligo(phenylene ethynylene) molecules, scanning tunneling microscopy, density functional theory, carbon nanotubes, electron-beam-induced oxidation and helium-ion microscopy and sputtering.

2 Materials and Methods

2.1 Oligo(phenylene ethynylene)s

Historical Background

Oligo(phenylene ethynylene)s (OPEs) are conjugated organic oligomers of successive groups of phenylene and ethynylene. For an explanation of the nomenclature, see Fig. 2.1. OPEs have been discussed as potential molecular wires since the beginnings of molecular electronics in the early nineties [16, 17]. Their electronic structure is assumed to be a conjugated π-electron system along the axis of the molecule.

Interest in OPE molecules increased when Tour et al. reported on negative differential resistance (NDR) of a self-assembled monolayer (SAM) of amino-nitro substituted OPE molecules between two gold electrodes [18]. The NDR effect was attributed to a reversible electrochemical redox reaction of the molecule [19]. The synthetic chemistry to fabricate such molecular wires has been extensively reported on by the same group [20], and their steady involvement in the field has led to OPE wires being referred to as *Tour wires*. Chemically similar conjugated organic oligomers, such as oligo(thiophene ethynylene)s and oligo(phenylene vinylene)s are not the focus of this work, but nonetheless they shall be mentioned here as possible alternatives to OPE molecules [5]. Compared to them, OPE molecules have the advantage of a rigid backbone.

Conductivity of OPEs

Although proposed as replacements for wires between active components of future electronics, surprisingly few is still known about the electronic properties of single

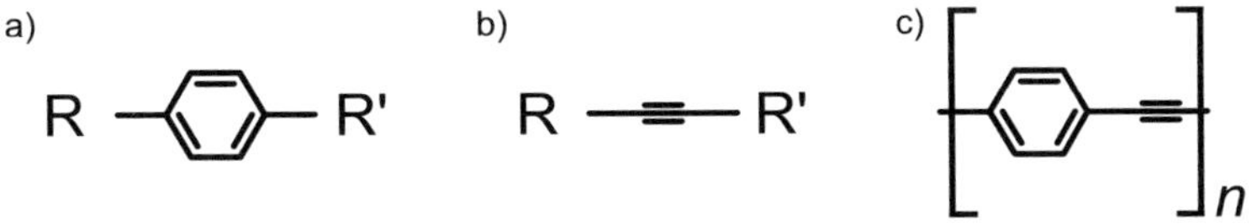

a) R–(phenylene)–R'   b) R–≡–R'   c) [–(phenylene)–≡–]$_n$

Figure 2.1: a) Phenylene group b) Ethynylene group c) n subunits of an oligo(phenylene ethynylene) (OPE).

OPE molecules. To isolate single OPE molecules from each other, they were embedded in SAMs of n-alkanethiolates on a gold surface. However, it could only be determined qualitatively that their conductivity is higher than that of the surrounding alkanes of equal length [21, 22]. Statistical methods such as mechanically-controlled break junctions (MCBJs) [8] or scanning tunneling microscope break junctions have since been relied on for conductance measurements of OPE molecules. The results are often smeared out by effects of the molecular conformation on the conductance [23], as the measurement presents only an average over all of them.

STM break-junction data of a three-subunit OPE between two gold contacts in solution indicate a conductance of 13 nS (corresponding to $R \approx 77\,\mathrm{M\Omega}$) [24]. The addition of a nitro substituent in the central phenylene group reduced this to 6 nS ($R \approx 167\,\mathrm{M\Omega}$). The conductance of longer OPE molecules has only recently been the focus of research. Wang et al. reported STM break-junction measurements of the conductance of OPEs with up to seven phenylene groups and a length of ≈ 5 nm [25]. A transition in the transport mechanism from tunneling to hopping transport was observed at an OPE length of ≈ 2.75 nm. Longer OPE molecules have been synthesized with up to 16 subunits, but only optical spectroscopy of them was reported [26].

STM studies of OPEs

Scanning probe studies of OPE molecules in the literature are limited to self-assembled monolayers [21, 22, 27–31] and associated STM break-junction data, with most studies focusing on rod-like structures. While U-shaped OPEs have been synthesized as early as 2004 by Tour et al. [32], only recently the interest of the scientific community in shape-persistent macrocycles has sparked further research into complex geometries: Höger et al. have extensively reported on the synthesis and characterization of linear, kinked and shape-persistent macrocycles of OPE variants [33–35] and analyzed their varying self-assembly behavior with an STM. Scanning probe studies of single, isolated (i.e. not embedded in an organic SAM) OPE molecules have not been reported so far.

Theoretical Studies of OPEs

First theoretical-chemistry studies on OPE molecules aimed at explaining the observed NDR effect [36]. Density functional theory calculations of an OPE with three phenylene groups and amino-nitro substituents on the central ring revealed that the molecular orbitals change their localization upon charging the molecule. In a later theoretical study the rotation of the phenyl rings to each other, i.e., the molecular conformation, was confirmed to play a central role for the molecule's conduction as well [37]. Recently, the same effect was found for biphenyl molecules with a chemically fixed twist angle between the two phenyl rings in break junction measurements [23,

38]. The reported shapes of the lowest unoccupied molecular orbital (LUMO) and highest occupied molecular orbital (HOMO) depend on the side chains of the central subunits as well as on the torsion angle between them. For planar configurations (torsion angle 0°) without substituents, delocalized orbitals are observed for both the LUMO and the HOMO [36][1]. Further theoretical studies focused on the influence of terminal atoms [40] and of various substituents and an external electric field [39]. All of these studies were performed on molecules containing at most three phenylene groups in the gas phase. The influence of a surface or of hybridization with metallic contacts is crucial with regards to possible applications as a molecular wire but has not been considered in these calculations.

2.1.1 OPE-A and OPE-B

Recently, OPE molecules have been used to integrate functional molecular electronics devices contacted by carbon nanotubes (CNTs): A CNT-molecule-CNT junction was reported by Marquardt et al., where a central dye molecule, a naphtalene diimide chromophore, was extended on both sides by three-subunit OPEs serving as conductive bridges [14]. On both of its ends the molecule possessed phenanthrene anchor groups to allow π-π coupling to the sidewall of the CNT [41, 42]. The onset of electroluminescence of the chromophore core was observed at bias voltages larger than 4 V.

In this work, we study two similar variants of OPE molecules. As we also aimed for transport measurements of our molecules between carbon nanotube electrodes, the phenanthrene anchor on each end was maintained. However, the chromophore core was not inserted, as its steric alignment is believed to be perpendicular to the OPE rods, possibly impeding STM measurements. Two planar OPE molecules, depicted in Fig. 2.2, were finally selected. The one with five subunits will be called OPE-A throughout this work, while the other one will be referred to as OPE-B. OPE-A consists of five phenylene-ethynylene subunits, while OPE-B consists of two two-subunit OPEs joined in the center by a bi-ethynylene group. Two hexyl chains are attached at the phenylene groups of each subunit to promote solubility in organic solvents. While OPE-A resembles a straight wire, the integration of the bi-ethynylene group at its center makes OPE-B resemble a junction of two shorter wires. In the future, a functional group could be integrated at its center without much changing the wire's structure. Both molecules were synthesized in the group of Prof. Marcel Mayor at the University of Basel, by Dr. Thomas Eaton and Dr. David Muñoz Torres, by

[1] Seminario et al. do not explicitly plot the HOMO but instead repeat the calculation with the molecule in charge state $Q = -2$ (two electrons removed) and show the resulting LUMO. Because the original LUMO was delocalized over the whole length of the molecule, this measure changes the potential evenly and should yield almost the same orbital as the HOMO of the molecule in neutral charge state ($Q = 0$). Indeed, the reported shapes are identical to the HOMO and LUMO reported by Li et al. [39] for the neutral molecule in zero electric field.

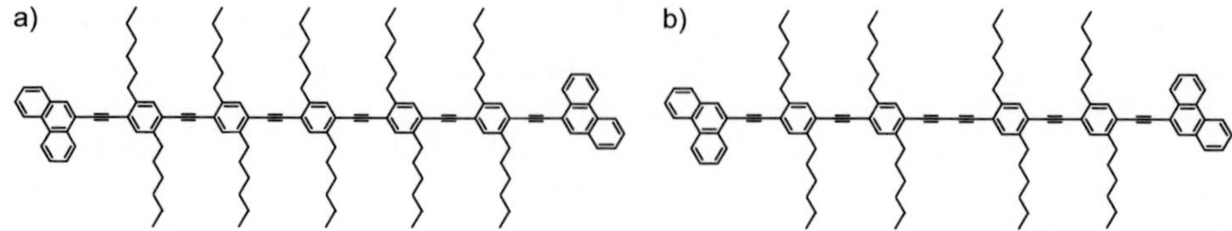

Figure 2.2: Structure of both oligo(phenylene ethynylene)s investigated in this work. a) OPE-A: $C_{130}H_{158}$ (1720.65 g/mol). b) OPE-B: $C_{112}H_{130}$ (1476.23 g/mol). The conjugated backbone consists of a) five phenylene-ethynylene subunits b) two pairs of two phenylene-ethynylene subunits connected by two ethynylene groups. At each end of both molecules, phenanthrene anchor groups provide the possibility of π-π coupling to the sidewall of a carbon nanotube. The hexyl chains on each side of a phenylene group promote solubility in organic solvents.

performing a series of acetylene protection and deprotection steps, similar to the molecule used by Grunder et al. [43]. Full synthetic details for OPE-A can be found in the supplementary material of [1]. OPE-A ($C_{130}H_{158}$) and OPE-B ($C_{112}H_{130}$) possess molecular weights of 1720.65 g/mol and 1476.23 g/mol, respectively.

Fluorescence Spectroscopy of OPE-A and OPE-B

A Varian Cary Eclipse Fluorescence Spectrophotometer was used to record fluorescence maps of solutions of both molecules in dichloromethane. Monochromated light with wavelengths from 250 nm to 600 nm was used to excite the molecules and the resulting fluorescence signal was spectrally analyzed. The background signal originating from the solvent was separately measured and subtracted from the data.

The fluorescence maps are shown in Fig. 2.3 and Fig. 2.4 for OPE-A and OPE-B, respectively. Both molecules show several pronounced emission peaks, with an onset around 420 nm. For OPE-A, a wider range of excitation wavelengths is able to excite this emission from the molecule. Horizontal slices of the data, for excitation wavelengths of 400 nm for OPE-A and 425 nm for OPE-B are shown in Fig. 2.5 and Fig. 2.6. In these spectra, four peaks were identified and fitted by Lorentzians. Under the assumption that the highest-energy peak arises from an electron-hole pair recombination between the HOMO and the LUMO level of the molecule, the HOMO-LUMO gaps of both molecules were extracted. The further peaks may be transitions between other levels of the molecule or phonon replica of the HOMO-LUMO transition.

A HOMO-LUMO gap of 2.95 eV (420 nm) was found for OPE-A and 2.90 eV (427 nm) for OPE-B. Lu et al. have performed ultraviolet-visible spectrophotometry on an OPE of similar length and report a value of ≈ 3 eV [25], in agreement with the fluorescence data.

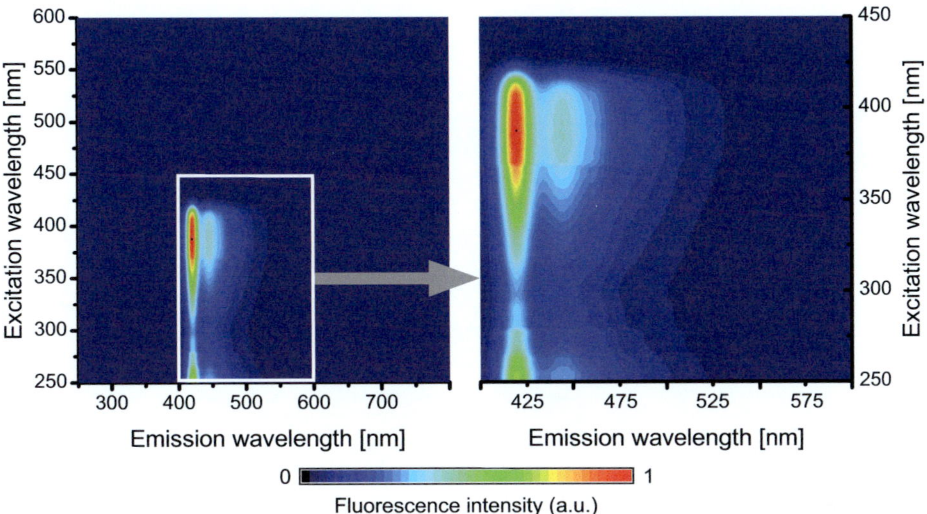

Figure 2.3: Fluorescence map of OPE-A in dichloromethane. The molecule shows an onset of emission at rougly 420 nm, with a wide range of excitation wavelengths able to excite this transition.

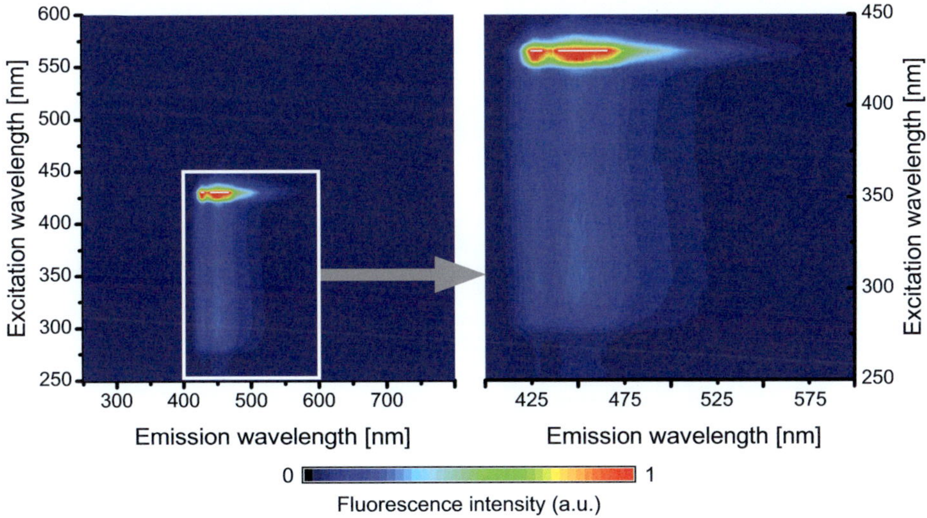

Figure 2.4: Fluorescence map of OPE-B in dichloromethane. Similar to OPE-A, an onset of emission at rougly 420 nm is observed, but with a narrower excitation wavelength window.

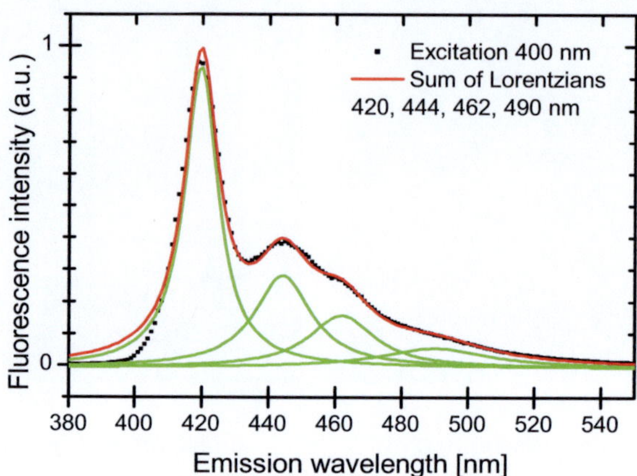

Figure 2.5: Horizontal slice through the fluorescence map of OPE-A in Fig. 2.3 for an excitation wavelength of 400 nm. Assuming the highest-energy transition arises from the HOMO-LUMO gap, an energy gap of 2.95 eV can be extracted from the data.

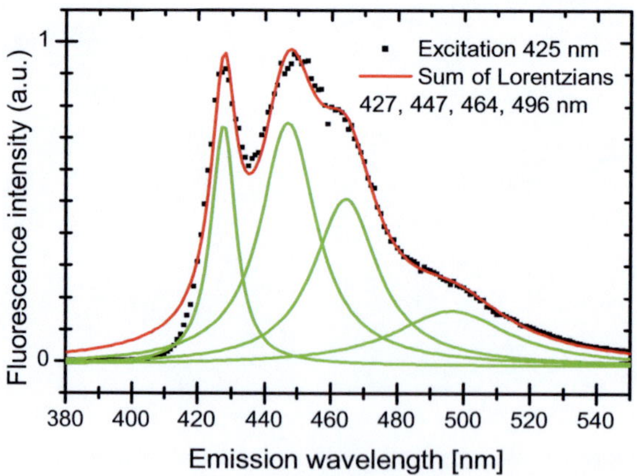

Figure 2.6: Horizontal slice through the fluorescence map of OPE-B in Fig. 2.4 for an excitation wavelength of 425 nm. Assuming the highest-energy transition arises from the HOMO-LUMO gap, an energy gap of 2.90 eV can be extracted from the data.

2.2 Scanning Tunneling Microscopy

Historical and Theoretical Background

The scanning tunneling microscope (STM), as first demonstrated by Binnig and Rohrer in 1979 [9, 10], is based on the vacuum tunneling of electrons between two closely spaced electrodes. The theory of STM in this chapter is reproduced from books by R. Wiesendanger [44] and D. Bonnell [45]. A voltage is applied between a sharp metal tip and a conductive surface, which are scanned relative to each other in close proximity. The resulting steady-state tunneling current I between these two reservoirs can be described in the framework of J. Bardeen [46] by

$$I \propto \sum_{\mu,\nu} \int_{-\infty}^{\infty} (f(E_\mathrm{F}-eV+\epsilon)-f(E_\mathrm{F}+\epsilon))\rho_\mathrm{tip}(E_\mathrm{F}+\epsilon)\rho_\mathrm{sample}(E_\mathrm{F}-eV+\epsilon)|M_{\mu\nu}|^2 \mathrm{d}\epsilon \quad (2.1)$$

where f denotes the Fermi distribution, E_F the Fermi level and ρ_tip and ρ_sample the local density of states of the tip and sample, respectively. The matrix element $M_{\mu\nu}$ contains their wave function overlap:

$$M_{\mu\nu} = -\frac{\hbar^2}{2m} \int (\psi_\mu^* \nabla \Psi_\nu - \Psi_\nu \nabla \psi_\mu^*) \mathrm{d}\vec{S} \quad (2.2)$$

The expression is integrated over the surface region separating the two electrodes. $M_{\mu\nu}$ thus depends on the geometry and the electronic properties of tip and sample. At low temperatures the Fermi distributions can be replaced by step functions and, using the transition probability $T(\epsilon, eV) = \sum_{\mu,\nu} |M_{\mu\nu}|^2$ the integral written as:

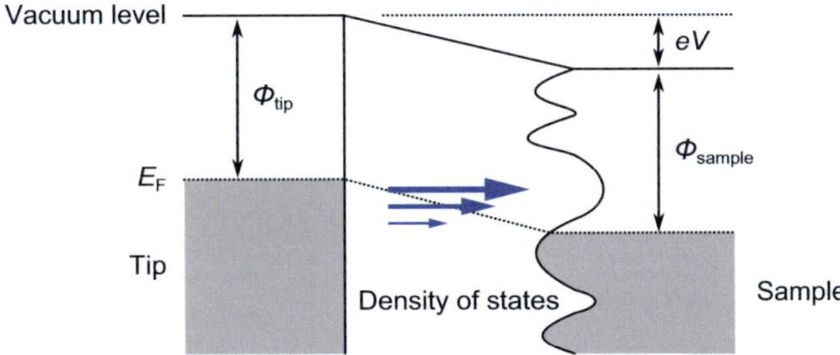

Figure 2.7: Tunneling between tip and sample with a positive bias voltage V probes the unoccupied states of the sample, given that the density of states of the tip is flat. ϕ_sample and ϕ_tip are the work functions of sample and tip, respectively. Electrons at the Fermi level E_F contribute more strongly to the tunneling current than electrons at lower energies.

$$I = \int_0^{eV} \rho_{\text{sample}}(E_\text{F} - eV + \epsilon)\rho_{\text{tip}}(E_\text{F} + \epsilon)T(\epsilon, eV)\mathrm{d}\epsilon \tag{2.3}$$

Under the assumption of a spherical s-type tip wave function, Tersoff and Hamann evaluated the matrix element [47, 48] and found an exponential dependence of the tunneling conductance on the distance between tip and sample. This exponential dependence leads to the high resolution of the STM. Using the Wentzel-Kramers-Brillouin (WKB) approximation, one can approximate the tunneling current through a trapezoidal barrier with:

$$T = \exp\left(-\frac{2d\sqrt{2m_\text{e}}}{\hbar}\sqrt{\frac{\phi_{\text{sample}} + \phi_{\text{tip}} + eV}{2} - \epsilon}\right) \tag{2.4}$$

Here m_e denotes the effective electron mass, d the distance between sample and tip, and ϕ_{sample} and ϕ_{tip} the respective work functions. Assuming a constant distance and a low bias voltage, so that the change in the transition probability is negligible, one can see that the differential conductance is directly proportional to the local density of states of the sample, given that the density of states of the tip is flat:

$$I \propto \int_0^{eV} \rho_{\text{sample}}(E_\text{F} - eV + \epsilon)\mathrm{d}\epsilon \tag{2.5}$$

$$\frac{\mathrm{d}I}{\mathrm{d}V} \propto \rho_{\text{sample}}(E_\text{F} - eV) \tag{2.6}$$

The differential conductance can be normalized by the total conductivity to make it less dependent on the exponential relation of the tunneling current to tip-sample separation and bias voltage [49].

Realization of an STM

Binnig and Rohrer were awarded the Nobel prize in physics in 1986 for the construction of the first scanning tunneling microscope, see Fig. 2.8 for a sketch of their setup. Since then the basic principle has not changed. A sharp metal tip is mounted on a XYZ-piezo scanner and brought close to a surface. A voltage is applied between the tip and surface and the tunneling current between the two is monitored.

In this work, two different operating modes of the STM are used:

- In **constant-current mode** a feedback loop controls the distance between tip and sample via the Z-piezo voltage trying to keep the tunneling current constant. This voltage signal corresponds to the apparent height of the surface and is recorded at each pixel to generate an image of the surface.

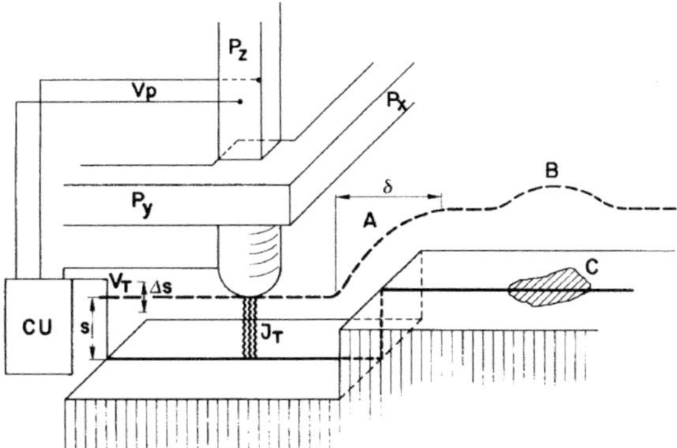

Figure 2.8: Sketch of Binnig and Rohrer's scanning tunneling microscope. A control unit (CU) applies a voltage V_{T} between tip and sample and measures the tunneling current J_{T}. A feedback loop applies a voltage V_{p} on the Z-piezo P_{Z} and tries to keep the tunneling current constant. The surface is scanned by controlling the P_{X} and P_{Y}-piezos. The Z-piezo voltage can be interpreted as a height signal and is used to construct an image of the surface. Reproduced from [10].

- In **constant-height mode** the feedback loop is disabled. The tip is positioned at a fixed distance from the surface and the tunneling current is recorded while scanning the surface.

Typically, constant-height mode allows faster imaging, because the scan speed does not have to accommodate for the regulation delay of the feedback loop. However, it can only be used on known "flat" areas of a sample, because the tip must not come too close to the surface or even crash into it. Drift and creep of the Z-piezo or drift of the sample can also lead to the tip crashing into the surface, which must be avoided. Hence, constant-current mode is typically used for "topographic" images.

For spectroscopic measurements, the tip is placed above a region of interest and the feedback loop is disabled. Then, the voltage is ramped while the current I is recorded. The differential conductance $\mathrm{d}I/\mathrm{d}V$ can be directly measured by a lock-in amplifier (LIA): A small high-frequency modulation is added to the bias voltage, with the frequency high enough to not disturb the normal current (I) measurement. A LIA then multiplies the I signal with the original modulation and integrates the resulting signal over a couple of cycles. When the phase shift between the modulation of the original signal and the recovered signal is zero or $180\,^{\circ}$, the output signal is maximized and is proportional to the slope of the I signal at a given bias voltage. Components of the I signal that are out of phase or at other frequencies are attenuated, which is why LIAs are known as *phase-sensitive detectors*. They allow measurements in noisy environments because they are essentially band-pass filters and only susceptible to noise around their oscillation frequency. Typical devices contain an adjustable phase shifter for control of the phase delay of one of the signals before multiplication.

STMs can operate in air or in vacuum. Typically, vacuum conditions are preferred, as surfaces accumulate dirt in atmospheric conditions or even oxidize immediately. Furthermore, vacuum systems allow measurements at cryogenic temperatures (avoiding condensation) as well as high temperatures (avoiding oxidation, if not of the surface, then of surrounding materials).

In this work, a liquid-helium cooled STM in an ultra-high vacuum was used for all measurements. The home-built STM was designed by Prof. Dr. Wulf Wulfhekel and Dr. Maya Lukas and jointly assembled in our research group. As STM tips a tungsten wire prepared by the method described in [50] was used.

2.2.1 Atomic Resolution Imaging

Since its invention, the STM has been capable of resolving surfaces at atomic resolution. However, atomic resolution imaging of individual adsorbed molecules seemed to be out of reach, because the topographic contrast of the STM is based on the local density of electronic states, i.e., the molecular orbitals, which need not be correlated with a molecule's chemical structure. Indeed, on conductive surfaces the molecular orbitals can hybridize with the surface states and thus only a convolution of many orbitals can be imaged, smearing out the details of individual orbitals. Repp et al. showed in 2005 that a thin insulating layer, e.g., a few layers of NaCl on copper, inhibit this hybridization and allow a clear image of the HOMO and the LUMO of a pentacene molecule to be recorded [51]. Using this technique, the HOMO and LUMO of naphthalocyanine and methylterrylenes molecules could be directly imaged [52, 53]. The lateral resolution of the images could be improved by modifying the STM tip to end in a carbon monoxide molecule or an organic molecule itself [51, 54].

Recently, both non-contact atomic force microscopy (ncAFM) [55, 56] and scanning tunneling hydrogen microscopy [57] (STHM) have been reported to achieve atomic resolution of single organic molecules, i.e., direct images of the chemical structure. In both cases a modification of the tip is critical: For ncAFM, a carbon monoxide molecule is picked up, whereas for STHM a hydrogen atom is brought between tip and molecule. The underlying contrast mechanism may be Pauli repulsion in both cases [55, 58]. So far, STHM data has only been reported for 3,4,9,10-perylene-tetracarboxylic dianhydride (PTCDA) monolayers and pentacene molecules on Au(111) [57–59]. For STHM on PTCDA, a favorable interaction of a hydrogen atom at the apex of the STM tip with the molecule has been proposed as the cause of the resolution enhancement [60].

2.2.2 The Au(111) Surface

To image molecules with an STM, they need to be placed on a conductive, atomically clean and flat surface. Hence, for STM measurements at cryogenic temperatures, a

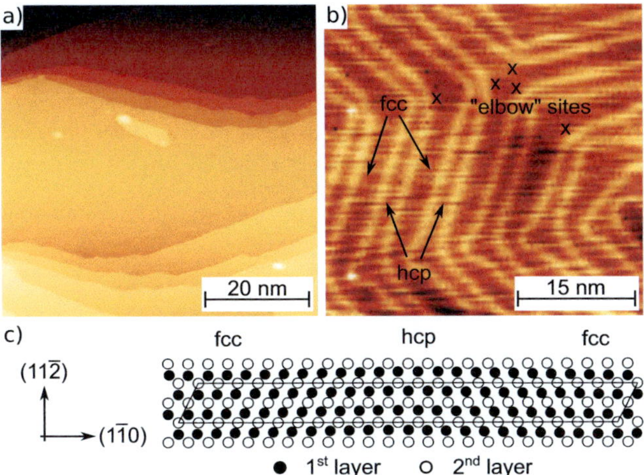

Figure 2.9: A clean Au(111) surface. a) The size of the terraces depends mainly on the miscut angle of the crystal and its preparation. b) The herringbone reconstruction. Transition regions appear brighter than the small hcp and larger fcc regions. Examples of "elbow" sites are indicated. Domains of the reconstruction meet at an angle of 120°. c) Top view onto the first two layers of the Au(111) reconstruction. The unit cell is indicated. Adapted from [69]. STM parameters: a) $I = 50\,\text{pA}$, $V = 1.8\,\text{V}$ b) $I = 200\,\text{pA}$, $V = 1.5\,\text{V}$.

gold single crystal was used in this work. The surface had been cut perpendicular to the (111)-direction to a precision better than 0.1°. The Au(111) surface exhibits a surface reconstruction, in which the topmost layer is compressed to contain more atoms per layer than the bulk material [61]. The $22 \times \sqrt{3}$ unit cell contains 46 atoms, which occupy 44 bulk-like positions [62, 63]. This leads to a varying stacking order from hexagonal close packing (hcp) to face-centered cubic packing (fcc) across the surface, with transition regions in between.

Clean gold crystal surfaces can be prepared by several cycles of argon-ion sputtering and annealing at 500 °C in ultra-high vacuum [64, 65]. An STM image of such a clean gold surface can be seen in Fig. 2.9. The size of the (111)-direction terraces mainly depends on the miscut angle of the crystal and the preparation. The step edges have a height of 2.35 Å. A closer view of one terrace can be seen in Fig. 2.9b, where the reconstruction is clearly visible. The bright lines are the transition regions between fcc and hcp stacking. The fcc regions are larger in size than the hcp regions [61]. Domains of the reconstruction have an angle of 120° to each other and distribute the elastic stress evenly across the surface [62, 66]. This reconstruction is known as the *herringbone reconstruction* because of its characteristic appearance. Scanning tunneling spectroscopy [67] as well as potential landscape mapping [68] agree on the electron binding energies of the gold surface: At the "elbow" sites it is weakest, followed by the fcc, then the hcp, then the transition regions.

2.3 Density Functional Theory

The electronic structure of a quantum system is governed by the many-particle Schrödinger equation. For an N-electron system in the Born-Oppenheimer approximation, where the position of the positively charged nuclei are seen to be fixed, it reads:

$$\hat{H}\Psi = \left(\underbrace{\sum_{i}^{N} \left(-\frac{\hbar^2}{2m_i} \nabla_i^2 \right)}_{\hat{T}} + \underbrace{\sum_{i}^{N} V(\vec{r}_i)}_{\hat{V}} + \underbrace{\sum_{i<j}^{N} U(\vec{r}_i, \vec{r}_j)}_{\hat{U}} \right) \Psi = E\Psi \tag{2.7}$$

$\hat{T}$ is the kinetic energy, $\hat{V}$ the potential of the positively charged nuclei and $\hat{U}$ the electron-electron interaction energy. The wave function Ψ depends on $3N$ spatial coordinates, which causes the computational effort to increase quickly for systems with many electrons. Density functional theory (DFT) is an alternative approach to directly solving the Schrödinger equation to find the ground state of such a system. The Nobel Prize in Chemistry 1998 was awarded to Walter Kohn and John Pople for their work on DFT and computational methods in quantum chemistry. The theory of DFT in this chapter is reproduced from Kohn's Nobel lecture [70] and from a review by K. Capelle [71]. DFT is based on the Hohenberg-Kohn theorem [72], which states that the ground-state Ψ_0 of an N-electron system can be described in terms of its electron density $n_0(\vec{r})$:

$$n_0(\vec{r}) = N \iiint_i \mathrm{d}^{3N} r_i \Psi_0^* \Psi_0 \tag{2.8}$$

The ground-state wave function hence becomes a functional of its electron density:

$$\Psi_0 = \Psi[n_0] \tag{2.9}$$

To find the ground state, its energy has to be minimized, given by:

$$E_0 = \left\langle \Psi[n_0] | \hat{T} + \hat{V} + \hat{U} | \Psi[n_0] \right\rangle \tag{2.10}$$

To do this, one solves for N non-interacting single-electron wave functions ϕ_i, satisfying the Kohn-Sham equations [73]:

$$\left(-\frac{\hbar^2}{2m_i} \nabla^2 + V_{\text{eff}}(\vec{r}) \right) \phi_i(\vec{r}) = \epsilon_i \phi_i(\vec{r}) \tag{2.11}$$

with the effective single-particle potential V_{eff} given as:

$$V_{\text{eff}}(\vec{r}) = V(\vec{r}) + \int \frac{e^2 n(\vec{r'})}{|\vec{r} - \vec{r'}|} \mathrm{d}^3 r' + V_{\text{XC}}(n(\vec{r})) \tag{2.12}$$

These single-electron wave functions $\phi_i(\vec{r})$ are the orbitals that reproduce the total electron density of the original many-body system:

$$n_0(\vec{r}) = \sum_i^N |\phi(\vec{r})|^2 \tag{2.13}$$

V_{XC} is called the exchange-correlation potential and includes all many-particle interaction effects. As the Kohn-Sham equations depend on the effective potential, which in its turn depends on the electron density given by the solutions of the Kohn-Sham equations, the equations are usually solved in an iterative, self-consistent manner.

With the exception of a free electron gas, typically the exact form of the exchange-correlation potential is not known. However, established methods (*functionals*) exist for approximating it, which reproduce the physical properties of various real-world systems in calculations well. Local methods use only the electron density (e.g. Local Density Approximation (LDA)), whereas gradient-corrected methods also depend on its gradient (e.g. Generalized Gradient Approximation (GGA)). Exchange- and correlation-energy functionals are typically named after their inventor(s). The two functionals used in this work are:

- the **BP86 (Becke-Perdew-1986)** functional, which combines the gradient-corrected exchange functional of Becke [74] with the gradient-corrected correlation functional of Perdew [75].
- the **B3LYP (Becke, three-parameter, Lee-Yang-Parr)** [76] functional, which is a so-called hybrid functional.

Hybrid functionals combine exchange energies which have been calculated in an exact manner (Hartree-Fock energies, E_{X}^{HF}) with exchange and correlation functionals from other sources. The B3LYP functional combines the exchange functional of Becke, $E_{\text{X}}^{\text{GGA}}$, with the correlation functional of Lee, Yang and Parr, $E_{\text{C}}^{\text{GGA}}$ [77]. The coefficients of the functionals are the three parameters:

$$E_{\text{XC}}^{\text{B3LYP}} = E_{\text{X}}^{\text{LDA}} + 0.2(E_{\text{X}}^{\text{HF}} - E_{\text{X}}^{\text{LDA}}) + 0.72(E_{\text{X}}^{\text{GGA}} - E_{\text{X}}^{\text{LDA}}) + 0.81(E_{\text{C}}^{\text{GGA}} - E_{\text{C}}^{\text{LDA}}) \tag{2.14}$$

$E_{\text{X}}^{\text{LDA}}$ and $E_{\text{C}}^{\text{LDA}}$ are local-density approximations of the exchange and correlation functionals of Vosko, Wilk and Nusair [78].

Often, molecular orbitals are constructed from a linear combination of atomic orbitals in DFT calculations. A basis set provides the parameters for the construction of atomic orbitals of each element out of primitive Gaussian orbital functions:

$$G^{\beta}_{i,j,k} = N x^i y^j z^k e^{-\beta r^2} \tag{2.15}$$

$$\text{with} \quad r^2 = x^2 + y^2 + z^2 \tag{2.16}$$

Two different basis sets are used for the calculations in this work:

- In the **6-31g*** [79] basis set, inner shell orbitals are constructed from a linear combination of six primitive Gaussians. The two valence orbitals are formed by a linear combination of three and one Gaussian orbitals, respectively, making it a so-called split-valence basis set. The X-YZg(*) notation was introduced by John Pople in 1971 [80]. The * indicates that a polarization function is added to the valence orbitals, allowing for an angular dependence (except for hydrogen atoms).
- The **def-SVP** basis is a standard basis set in the DFT community, defined for most of the elements of the periodic table. Being a split-valence basis set with polarization functions, it is similar to the 6-31g* basis set, however, it also contains polarization functions for hydrogen atoms. The basis set parameters were taken from Schäfer, Horn and Ahlrichs [81] for hydrogen and carbon, and from Eichkorn, Weigend, Treutler and Ahlrichs [82] for gold.

2.4 Carbon Nanotubes

Historical and Theoretical Background

Carbon nanotubes (CNTs) were discovered in 1991 by Iijima in Japan [83]. They can be regarded as a rolled-up graphene ribbon, which is a single layer of graphite. Graphene was first hypothesized by Wallace in 1947 [84] and first observed by Hans-Peter Böhm in 1962 [85]. Its full potential was only realized much later, when it was rediscovered in 2004 by Geim and Novoselov [86]. Graphite layers are carbon atoms arranged in a hexagonal ("honeycomb") lattice. They can be described by a two-atom basis and the lattice vectors

$$\vec{a}_1 = \frac{a_{cc}}{2}\begin{pmatrix} 3 \\ \sqrt{3} \end{pmatrix} \quad \text{and} \quad \vec{a}_2 = \frac{a_{cc}}{2}\begin{pmatrix} 3 \\ -\sqrt{3} \end{pmatrix}, \tag{2.17}$$

with the carbon-carbon bond length $a_{cc} = 1.42\,\text{Å}$. Carbon has four valence orbitals, 2s and $2p_{x,y,z}$. The orbitals in plane with the graphene layer are hybridized into sp^2-bonds. The p_z-orbital remains perpendicular the graphene sheet. The atomic structure of a single-walled nanotube (SWNT) can be fully described by a chiral vector and its length. The chiral vector defines the roll-up direction of the graphene sheet, and is therefore a linear combination of the graphene basis vectors:

$$\vec{C}_{ch} = n\vec{a}_1 + m\vec{a}_2. \tag{2.18}$$

The chiral vector runs along the circumference of a carbon nanotube, and its length is related to the diameter of a tube by:

$$d = \frac{|\vec{C}_{ch}|}{\pi} = \frac{\sqrt{3}a_{cc}}{\pi}\sqrt{n^2 + nm + m^2}. \tag{2.19}$$

One specific type of nanotube chirality is referred to by its (n,m) tupel. $(n,0)$ tubes are known as zig-zag tubes and (n,n) tubes as armchair tubes, because of the resemblance of the shortest possible path following nearest-neighbor atoms along the circumference. See Fig. 2.10 for a graphical representation of the chiral vector on a sheet of graphene and atomic structures of a zig-zag, an armchair and a chiral CNT.

To a first approximation, the electronic properties of CNTs can be derived from the electronic properties of graphene. Fig. 2.11 shows the calculated band structure of graphene from a tight-binding approach. The bonding and antibonding σ-bands are separated by a large energy gap and lie far away from the Fermi level. However, the bonding and antibonding π-bands cross the Fermi level at the K-point. The Fermi

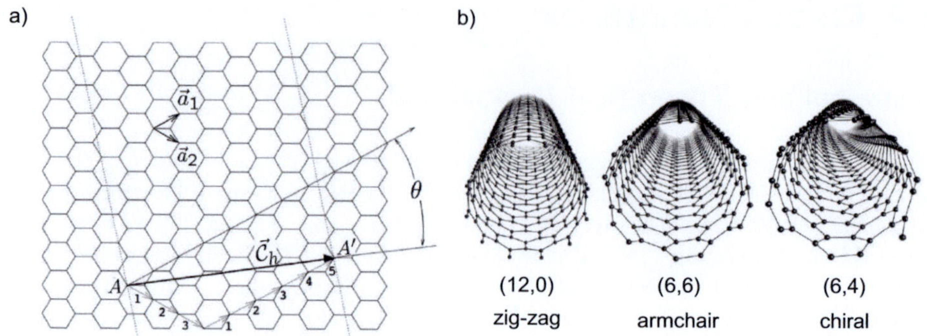

Figure 2.10: a) Illustration of the roll-up vector (chiral vector) of a (5,3) CNT. (n,m) denotes multiples of the graphene basis vectors. The chiral angle θ between $\vec{C}_{\mathrm{ch}}$ and $\vec{a}_1$ is also indicated. b) Atomic structure of a zig-zag ($\theta = 0$), armchair($\theta = 30\,°$) and a chiral nanotube. Adapted from [87].

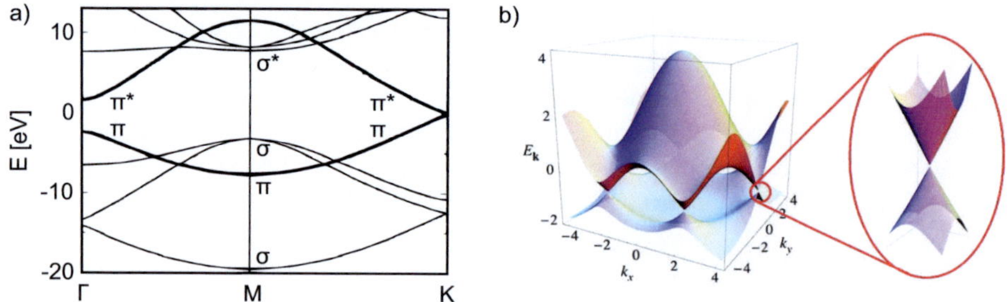

Figure 2.11: a) Band structure of graphene. Bonding and antibonding σ^*-orbitals are separated by a large energy gap. Bonding π- and antibonding π^*-orbitals cross at the K-point. Adapted from [87]. b) 3D view of the band structure, highlighting the Dirac cone at a K-point. The dispersion relation is linear around this point. From [89].

surface thus consists of the six K-points at the corners of the first Brillouin zone, making Graphene a semimetal. The dispersion relation is linear around the K-point, i.e. it forms a so-called Dirac cone because the charge carriers behave like massless Dirac fermions. This leads to a number of interesting electronic and optical properties, for a review, see e.g., Bonaccorso et al. [88].

To derive the band structure of CNTs, the zone-folding approximation is used: The allowed k-vectors along the circumference of the tube are confined and have to fulfill the Bloch condition [87]:

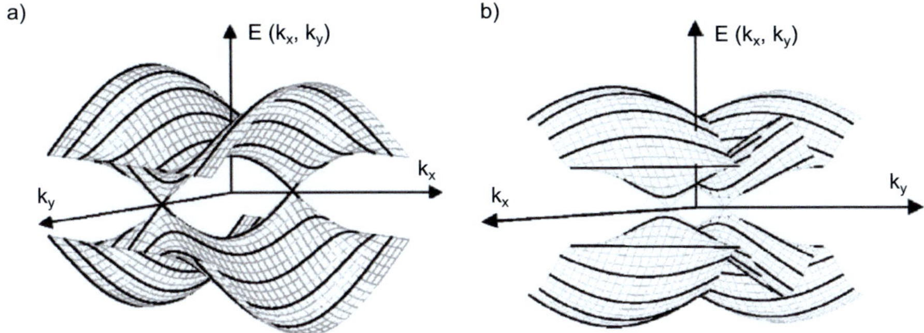

Figure 2.12: Graphene dispersion relation near the K-point overlaid with the allowed states for a) a metallic carbon nanotube and b) a semiconducting carbon nanotube as derived from the zone-folding model. Adapted from [87].

$$\Psi_{\vec{k}}(\vec{r} + \vec{C}_{\text{ch}}) = e^{\imath \vec{k}\cdot\vec{C}_{\text{ch}}} \Psi_{\vec{k}}(\vec{r}) = \Psi_{\vec{k}}(\vec{r}) \tag{2.20}$$

$$\Leftrightarrow \vec{k} \cdot \vec{C}_{\text{ch}} = 2\pi l \tag{2.21}$$

The projection of the allowed k-vectors for a CNT onto the Brillouin zone of graphene gives the band structure as slices through the original graphene band structure. If the K-points lie on one of these lines, the nanotube is metallic, otherwise it is semiconducting. See Fig. 2.12 for this projection for a metallic and a semiconducting CNT. If we insert the coordinates of one of the K-points into Eq. 2.21, we get

$$\left(0, \frac{2\pi}{a_{\text{cc}}} \frac{2}{3\sqrt{3}}\right) \cdot (n\vec{a_1} + m\vec{a_2}) = 2\pi l \tag{2.22}$$

$$\Leftrightarrow \frac{2\pi}{3}(n - m) = 2\pi l \tag{2.23}$$

This is fulfilled for $n - m = 3l$. Consequently, two thirds of all possible nanotubes are semiconducting and one third is metallic. Common synthesis methods always yield a mixture of different chiralities.

2.4.1 Synthesis of Carbon Nanotubes

Typical methods of carbon nanotube synthesis include chemical vapor deposition and methods that rely on the sublimation of carbon atoms in an inert gas atmosphere or in vacuum:

- **Chemical vapor deposition** growth of carbon nanotubes is based on the decomposition of a carbon precursor into nanotubes with the help of a catalyst [90]. Typical metal catalysts are iron, nickel, cobalt or molybdenum and the growth can happen either on a surface or in a gas atmosphere at temperatures $\leq 1000\,°C$. A typical precursor gas is methane, but others can be used, too.
- The high temperature of an **arc discharge** between two graphite electrodes in an inert gas atmosphere can lead to the sublimation of graphite. The growth of CNTs then starts at catalyst particles that were mixed into the graphite electrodes [91].
- For **pulsed laser vaporization** [92], a graphite target with catalysts is hit by an energetic laser pulse. Carbon sublimates and eventually forms carbon nanotubes at catalyst particles. All reaction products are collected on a filter paper downstream of an inert gas flow.

The choice of catalyst influences the growth speed, the required temperature and the resulting diameter and length distributions of the carbon nanotubes. However, not only single-walled carbon nanotubes are formed. Multi-walled tubes, fullerenes, graphene flakes and amorphous carbon are produced as well and have to be removed. For a detailed review of common synthesis techniques see [93].

2.4.2 Preparation of Nanotube Suspensions

For the fabrication of nanogaps by electron-beam-induced oxidation, the carbon nanotube suspensions were prepared by Dr. Frank Hennrich. The CNTs were produced by pulsed laser vaporization (PLV) as described in the previous section. To filter out amorphous carbon and other large impurities, the soot caught in the filter paper was first washed in dimethylformamide three times and then centrifuged. The supernatant was collected and the nanotubes were subsequently dispersed in water by sonication in 1 weight/volume% (w/v %) sodium cholate and then sorted by column chromatography, by the method described in [11]. Fractions of metallic single-walled carbon nanotubes (mSWNTs) with a diameter distribution of (1.2 ± 0.2) nm and a length of (1 ± 0.5) µm were selected.

The carbon nanotubes used for helium-ion sputtering of nanogaps were prepared by Dr. Benjamin S. Flavel. CNT material from PLV was sorted by S-200 gel filtration [94] and further purified by density-gradient ultracentrifugation (DGU). Typically 10 mg of raw CNT material was suspended in 15 ml H_2O with 1 wt.% of sodium dodecyl sulfate (SDS) using a tip sonicator (Bandelin, 200 W maximum power, 20 kHz, 100 ms pulses) for 2 h at $\approx 20\,\%$ power for an initial suspension. During sonication the suspension was cooled by a 500 ml water bath. The resulting dispersion was then centrifuged with $\approx$100.000 g for 1.5 h and carefully decanted from the pellet which was formed during centrifugation. The centrifuged CNT suspension was used as the starting suspension for gel filtration fractionation. Gel filtration was performed in a glass column of 20 cm length and 2 cm inner diameter. After filling the glass column with the filtration medium, the gel was slightly compressed to yield a final height of ≈ 14 cm. For the separation, ≈ 10 ml of the initial suspension was applied to the top of the column and subsequently a solution of 1 wt. % SDS in H_2O as eluant was pushed through the column with compressed air by applying sufficient pressure to ensure a flow of ≈ 1 ml/min. After ≈ 10 ml of this eluant had been added most of the mSWNTs had moved through the column, whereas the semiconducting CNTs remained trapped in the upper part of the gel [94]. The mSWNTs were collected and used for DGU, to remove defected mSWNTs and any additional carbonaceous species present. Ultracentrifugation was performed in 20 wt.% iodixanol and 1 wt.% SDS in H_2O. mSWNTs with a diameter distribution of (1.2 ± 0.2) nm were selected.

2.4.3 Dielectrophoresis of CNTs

Dielectrophoresis was first described in 1951 [95] and is the motion of a dielectric particle in an inhomogeneous electric field. It is based on induced multipolar moments and hence does not require the particle to be charged. The magnitude of the force depends on the size, shape and orientation of a particle, its dielectric environment and its polarizability.

If a nanotube is considered as a rod-like particle, it will orient itself along the field lines in a given inhomogeneous field because of the dielectrophoretic torque it experiences. The time-averaged dielectrophoretic force perpendicular to these field lines acting on a nanotube in suspension can be written as [96]:

$$\left\langle \vec{F}_{\mathrm{DEP}} \right\rangle = \frac{\pi d^2 l}{8} \epsilon_l \Re \left(\frac{\epsilon_t^* - \epsilon_l^*}{\epsilon_l^* - (\epsilon_t^* - \epsilon_l^*) L} \right) \nabla E^2 \tag{2.24}$$

Here, l is the length, d the diameter and L the depolarization factor of the nanotube. The latter is of the order of 10^{-5}. The term in brackets is called the Clausius-Mossotti function and contains all frequency-dependent parts of the dielectrophoretic force. ϵ_t^* and ϵ_l^* are the complex permittivities of the nanotube and the liquid, respectively:

$$\epsilon_{t,l}^* = \epsilon_{t,l} - i\frac{\sigma_{t,l}}{\omega} \tag{2.25}$$

For metallic carbon nanotubes, the dielectric constant and the conductivity are usually both higher than for the surrounding liquid. Hence, the tubes always experience a positive dielectrophoretic force, i.e. a force towards higher field strengths. For semiconducting nanotubes, a cross-over in the force direction depending on the frequency was found and dielectrophoresis was considered as a way of sorting nanotubes by

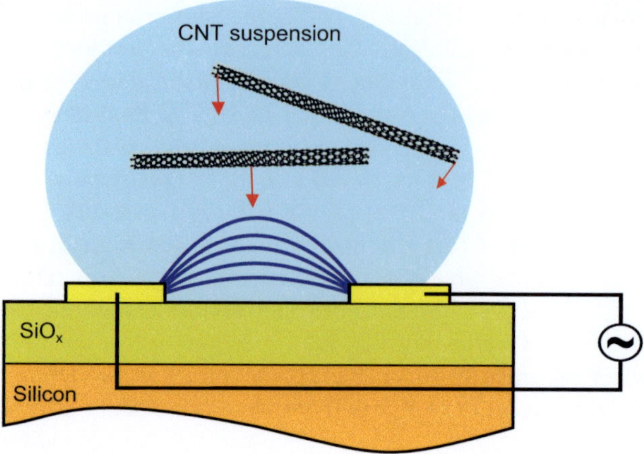

Figure 2.13: Dielectrophoresis for CNT device assembly. A drop of CNT solution is placed over electrodes which have an alternating electric field applied between them. Each tube experiences forces due to its polarizability and orients itself along the field lines. Metallic CNTs exhibit positive dielectrophoresis, meaning they will always be driven towards higher fields, and hence are placed on top of the contacts.

electronic type [97]. Nowadays, other sorting methods have proven more reliable and dielectrophoresis is mostly used as an assembly technique [98, 99], as it is in this work, too.

An alternating electric field is applied between two electrodes and a drop of nanotube solution is placed over them. Eventually, the nanotubes orient themselves in parallel to the electrodes and place themselves between them, if the dimensions of the electrodes and the deposition parameters (voltage, frequency) are chosen appropriately [100].

2.4.4 Electron-Beam Lithography

Electron-beam lithography (EBL) was used for the fabrication of metallic contacts on several types of substrates. It is an established method and therefore it will be described only shortly. An illustration of the processing steps is shown in Fig. 2.14.

A sample is first spin-coated with a suitable resist. Poly(methyl methacrylate) (PMMA) is an industry standard and was used in different formulations. The designation PMMA 950K A4.5, e.g., indicates that polymer chains with an average weight of 950 kDa are dissolved in anisol with a concentration of 4.5 weight/volume,%. The weight (length) of the polymer chains influences the sensitivity, whereas the solvent and concentration influence the layer thickness. If not specified otherwise, PMMA 950K A4.5 was used. With a spin-coating speed of 5000 rpm, the resulting film thickness is $\approx$ 200 nm.

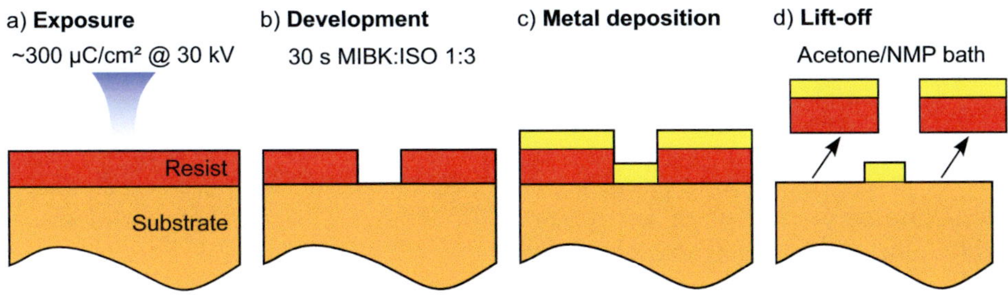

Figure 2.14: Process flow of electron-beam lithography. a) Electron-beam exposure of the resist layer. Typically, one or two layers of poly(methyl methacrylate) (PMMA) were used. b) Development of the resist in a 1:3 mixture of methyl isobutyl ketone (MIBK) in isopropanol. Exposed parts of the resist are removed. c) Metal is sputtered onto the sample, covering the resist and the exposed parts of the substrate. d) Acetone or n-methyl-2-pyrrolidone (NMP) are used to strip the resist layer together with the unwanted metal.

Photoresist layers are baked after spin-coating in a hot-air oven at 165 °C for 30 min to facilitate crosslinking of the polymer and to evaporate any solvent residue. For an electron exposure dose of $\approx 360\,\mu\mathrm{C/cm^2}$ at 30 kV, PMMA is a positive resist, i.e. its solubility in the developer increases with increasing dose. This means a positive (not inverted) pattern of the desired structure has to be exposed. After the exposure, a 1:3 mixture of methyl isobutyl ketone (MIBK) in isopropanol was used for the development. Samples are submerged in the mixture for 30 s, then rinsed with isopropanol and blow-dried. The developer removes the exposed part of the resist. Before further processing, samples are again baked in an oven for 30 min at 90 °C, to harden the photoresist and evaporate any solvent residue from the development.

To fabricate metallic structures, the sample is then coated with the desired metal. In our case, sputter deposition of metals is employed. In a vacuum chamber, an argon plasma is ignited and acts on a metal target to eject atoms or clusters from its surface, which are then deposited on the sample. This leads to a uniform metal coverage of both the unexposed resist and the substrate surface. To finally lift off the parts of the resist with the metal on top, acetone or n-methyl-2-pyrrolidone (NMP) are used. Both can be heated up to 60 ° (acetone) or 100 ° (NMP) to increase their effectiveness. When all undesired metal has come off the sample, it is again rinsed with isopropanol and blow-dried.

2.5 Electron-Beam-Induced Oxidation

Historical Background

Stewart noted in 1934 that metal surfaces inside vacuum tubes become discolored after exposure to electron or ion beams [101]. They are eventually covered with a carbon film, which is deposited from residual gas inside the vacuum tubes. To this day, hydrocarbon deposition on samples under ion/electron bombardment is a concern in electron microscopy. Christy published a quantitative analysis of the effect in 1960 [102]: He found that the deposition rate of polymer films under electron radiation in vacuum was proportional to the partial pressure of silicone pump oil in the chamber and the beam current density. In 1976, the first deliberate use of such a pattern as a mask for metal evaporation was reported [103], termed *contamination lithography* [104]. From then on, the advancements in charged particle beam induced deposition and etching were driven by the needs of the growing semiconductor industry. With the advent of focused ion-beam (FIB) instruments, valuable photomasks could be analyzed and repaired in-situ [105]. While the ion beam allows sputtering of material, the introduction of selected organometallic precursor gases into the microscope chamber enables local deposition of material [106, 107]. On the other hand, the introduction of reactive gases into the FIB chamber was found to increase sputtering rates for some materials [108]. Nowadays, gas-assisted deposition precursors and etching gases have been developed for many materials [109].

Deposition of material and etching activated by a charged-particle beam are similar processes: In both cases a precursor molecule is introduced into a vacuum chamber at a controlled rate. It covers the sample surface according to the Langmuir adsorption equation:

$$\Theta = \frac{\alpha P}{1 + \alpha P} \tag{2.26}$$

Θ denotes the fractional coverage of the surface, P is the pressure and α a factor that depends on the molecule's sticking coefficient and the temperature. A higher sticking coefficient leads to a higher surface coverage. A higher temperature typically leads to an increase of desorption and hence lower surface coverage. A charged particle then interacts with an adsorbed molecule. For deposition processes, the molecule undergoes a dissociation into a volatile and a non-volatile product. Non-volatile products remain on the surface and constitute the deposited material. Volatile products desorb and are pumped from the chamber. An illustration of electron-beam-induced etching is shown in Fig. 2.15. Reactive species are created by the interaction of the adsorbed gas molecules with charged particles. These species then react locally with the surface and form volatile compounds, which are eventually pumped from the chamber after desorbing.

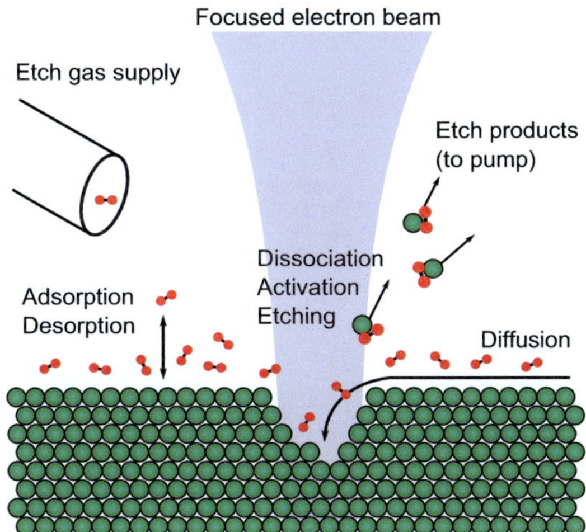

Figure 2.15: Scheme of electron-beam-induced etching. Precursor molecules cover the surface uniformly due to adsorption and thermal desorption. An electron beam activates the molecules and reactive species then combine with surface molecules, forming volatile compounds. These are eventually pumped from the system after desorption from the surface.

2.5.1 The Electron Microscope

With the invention of the electron microscope in 1932 by Knoll and Ruska [110] and subsequent refinements by von Ardenne [111], the interaction of charged particle beams with matter and surfaces of different kinds was studied. Already in 1940 von Ardenne realized that an electron beam with an acceleration voltage in the kilovolt range causes different kinds of low-energy electron radiation to leave a surface. What he called "nutzbare Strahlung" (usable radiation), later became known as type-1 secondary electrons (SE1). These are created by the primary beam at or near its point of entry. In addition, elastic scattering in a sample causes backscattered electrons (BSE) to escape the surface from a radius much larger than the SE1 electrons. Furthermore, backscattered electrons in turn create secondary electrons when passing the surface, which are known as type-2 secondary electrons (SE2). Von Ardenne called BSE and SE2 electrons "schädliche Strahlung" (harmful radiation), because these decreased his imaging resolution.

Secondary electrons are produced by inelastic collisions of the primary electrons with weakly bound outer-shell valence electrons (in semiconductors and insulators) or conduction electrons (in metals) [112]. The typical energy distribution of electrons emitted from a surface that is hit by primary electrons of energy E_{PE} is depicted in Fig. 2.17. Electrons with energies below 50 eV are usually called secondary electrons, whereas electrons with higher energies are identified as backscattered electrons.

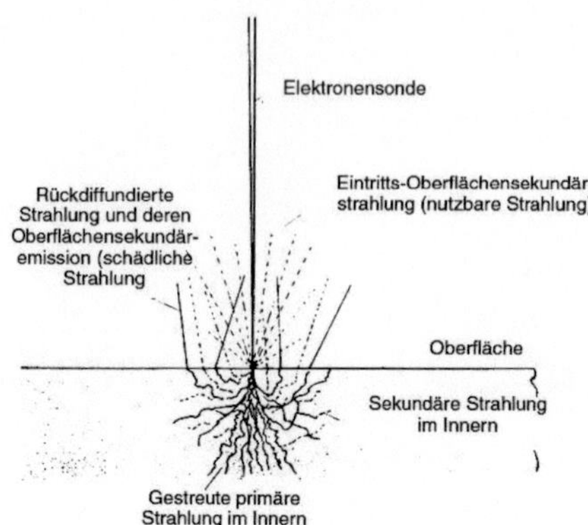

Figure 2.16: Types of secondary electrons created by an impinging charged particle beam. Type-1 secondary electrons (SE1) escape the surface from the primary point of impact of the beam. Backscattered electrons (BSE) stem from elastic scattering. Type-2 secondary electrons (SE2) are in turn created by backscattered electrons. BSE and SE2 electrons are not differentiated in this illustration. From [111].

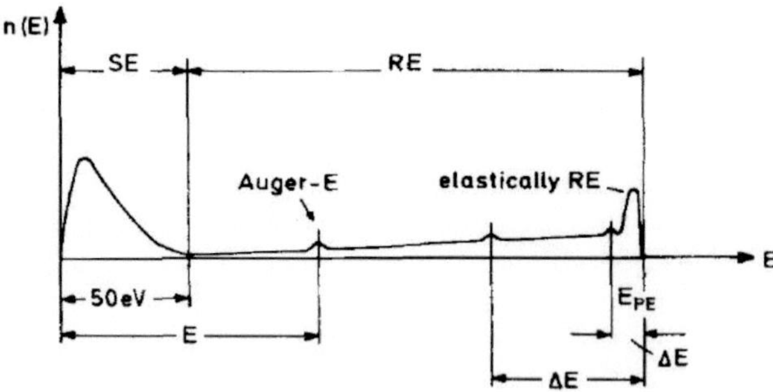

Figure 2.17: Energy distribution of electrons emitted from a surface impinged by primary electrons of energy E_{PE}. From [112].

Only the secondary electrons that escape the sample can be detected. The typical escape depth of secondary electrons in insulators is below 10 nm, down to a few nanometers for metals. In silicon oxide, the escape depth is around 8 nm [113].

2.5.2 Electron-Beam-Induced Oxidation of Carbon

For many molecules, dissociation and/or ionization cross section data in the gas phase are not readily available. To further complicate the matter, molecules on surfaces may have entirely different cross-sections due to catalytic effects of the surface (increasing the cross section) or fast surface relaxation processes that are not possible in the gas phase (decreasing the cross section). So, apart from the scarce data on gas-phase electron-impact cross-sections for relevant molecules, equivalent data for adsorbed molecules are also rare [109].

In this work, oxygen is used for EBIO of carbon nanotubes. It was reported that multi-walled carbon nanotubes and sputtered carbon thin films are etched by its reactive species [114–116]. Direct knock-on damage by primary-beam electrons can be excluded as the etching mechanism, as the threshold for carbon nanotubes is 86 keV [117], much higher than the electron energies in a typical SEM. Reactive species of molecular oxygen include oxygen molecules in an excited state, ionized oxygen atoms and ozone. Cross-sections for electron impact with oxygen molecules are in the range of 0.01 (dissociation) to 1 $Å^2$ (ionization) for electrons with kinetic energies in the range of ≈ 5 eV to several keV, see Fig. 2.18 for a review of the available data. From the data it is clear that secondary, backscattered as well as primary-beam electrons could be driving EBIO of carbonaceous compounds in an SEM.

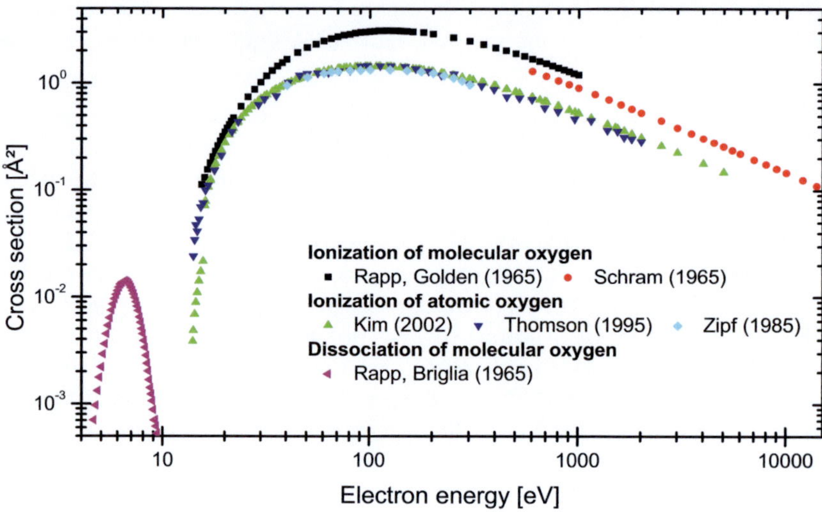

Figure 2.18: Cross-sections for ionization of atomic and molecular oxygen, as well as dissociaton of molecular oxygen by electron impact. Data from: Rapp, Golden (1965) [118], Schram (1965) [119], Zipf (1985) [120], Thompson (1995) [121], Kim (2002) [122], Rapp, Briglia (1965) [123].

2.6 Helium-Ion-Beam Lithography

2.6.1 Ion Beam - Sample Interaction

The Zeiss Orion Plus helium-ion microscope (HIM) used in this work allows imaging similar to a scanning electron microscope, except for a helium-ion beam being scanned over the sample. Ion optics focus and scan this beam across a sample surface, similar to the electron optics in an SEM. The ions are accelerated with fields up to 35 kV, similar to the field strengths used in SEMs. The interaction of the ion beam with a sample can be quite complex. As long as their kinetic energy is high (> 3 keV), the ions mostly lose their energy by inelastic collisions with the electron shells of atoms in the sample [124]. These collisions produce secondary electrons along the paths of the ions. Close to the surface, the secondary electrons can escape the sample and are detected in the chamber. The mass mismatch between helium ions and electrons minimizes the beam spread caused by these inelastic collisions, compared to the primary-electron beam of a SEM [125].

Due to the small convergence angle of the beam on the sample, the depth of field of HIM images is large compared to SEM images. The favorable shape of the interaction volume of helium ions compared to primary electrons in a conventional SEM leads to an improved resolution [124, 126, 127]. For a comparison of trajectories of 35 kV helium ions and 1 kV electrons with a similar surface sensitivity into a silicon oxide layer, see Fig. 2.19. The region where secondary electrons are generated close to the surface and eventually escape and reach the detector is smaller in the case of impinging helium ions.

Deeper in the sample, once the kinetic energy of the ions has been reduced, interactions with nuclei become dominant and the ions are eventually thermalized. Elastic collisions with the nuclei can lead to backscattered ions escaping from the sample, which can also be detected with an ion detector. Very thin samples also allow the transmission of ions. An illustration of an HIM is shown in Fig. 2.20.

The Gas Field Ion Source

The helium-ion source in the instrument relies on the principles of the field ion microscope, which was postulated by Müller in 1951 [129] and later realized in 1955 by him and Bahadur [130]. It was the first imaging technique with atomic resolution. A sharp metal tip is cooled to cryogenic temperatures in a low-pressure inert-gas atmosphere. A high voltage is then applied between the tip and an extractor plate. Neutral gas atoms are drawn to the tip surface via the polarization effect of the strong field. After thermalization, they are ionized by the tip and accelerated away by the field, perpendicularly to the surface. Detection of the ions on a photoplate shows a magnified view of the tip, with clearly distinguishable surface atoms.

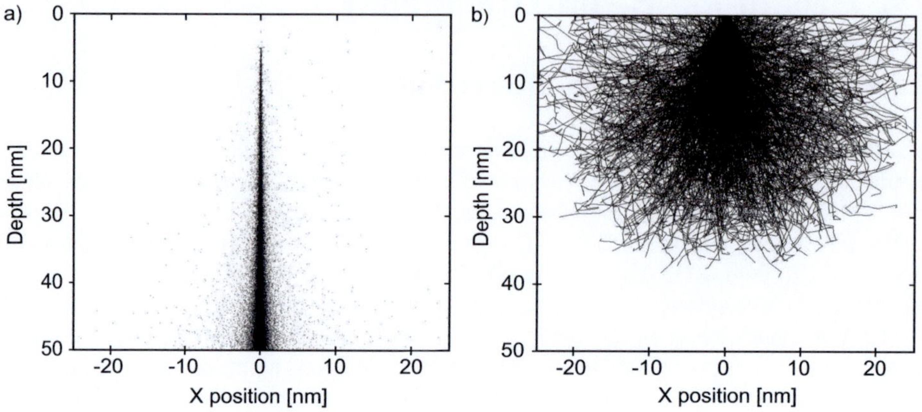

Figure 2.19: a) Near-surface trajectories of 20000 35 kV helium ions into silicon oxide. Plotted is the ion implantation distribution in the target. Calculated with SRIM (*Stopping and Range of Ions in Matter*) (http://www.srim.org). b) Near-surface trajectories of 1 kV electrons into silicon oxide. Calculated with CASINO (*monte CArlo SImulation of electroN trajectory in sOlids*).

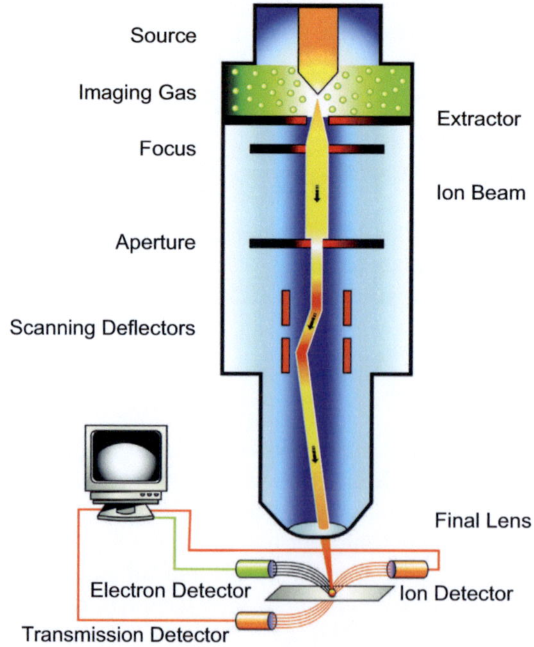

Figure 2.20: Scheme of a helium-ion microscope. Adapted from [128].

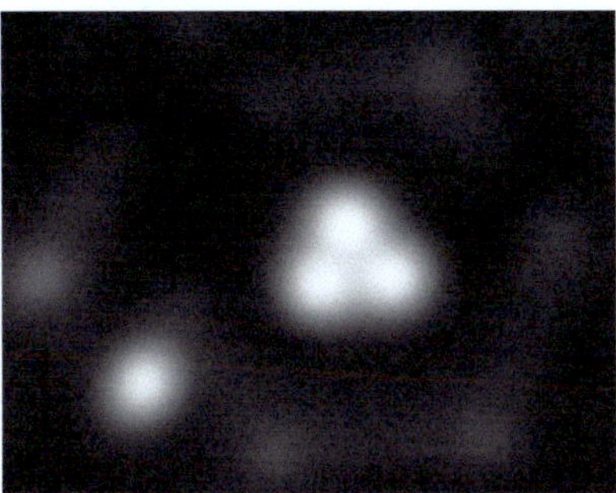

Figure 2.21: Field ion microscopy image of a gas field ion source in a helium-ion microscope. The emission from a trimer as well as from a defect atom in the tip's structure can be identified. The ion emission from the hexagonally ordered underlying layer is much weaker than of the atoms at the apex.

A proprietary process was invented by the ALIS corporation in 2006 to form the tip of a tungsten rod into a three-atom configuration [128]. This trimer configuration is sufficiently stable to be used as a gas field ion source for hours until it has to be rebuilt. A field ion image of the tip of such a source is shown in Fig. 2.21.

2.6.2 Ion-Beam Lithography

The lithographic capabilities of the helium-ion microscope are twofold:

- Due to the generation of secondary electrons by the focused ion beam, it can be used to expose conventional photoresists such as hydrogen silsesquioxane (HSQ) similar to an electron-beam-lithography system [131].
- Similar to the gallium ions in a conventional focused ion beam (FIB) instrument, a helium ion can transfer enough momentum to an individual atom of a sample by an elastic collision to eject it from the surface. This sputtering effect has so far been demonstrated on graphene [132–134], silicon nitride [135] and gold nanorods [136, 137]. Helium ions have a smaller sputtering yield than gallium ions of similar energies, enabling the fabrication of fine details on sensitive samples. Also, sample contamination by helium-ion implantation is usually less problematic than with gallium ions [138].

A theoretical study of ion sputtering of graphene using molecular dynamics simulation was recently published by Lehtinen et al. [139]. Probabilities for various defect types in a graphene sheet were simulated depending on the angle of ion incidence. For helium ions with a kinetic energy of 30 keV, single-vacancy type defects (with the missing atoms removed from the sheet) were most likely for an angle of incidence of $\approx 62\,°$. In principle, the surface of a carbon nanotube can be regarded as a curved

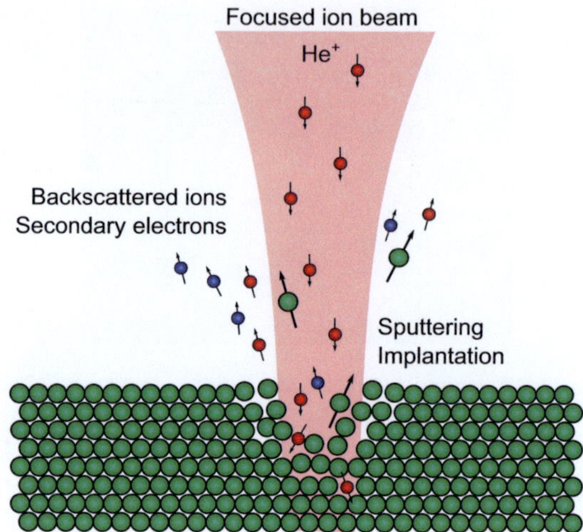

Figure 2.22: Scheme of helium-ion-beam lithography. The focused ion beam can sputter material off a surface. In normal imaging mode, the ion beam is scanned across the surface and secondary electrons and/or backscattered ions are detected to generate an image.

graphene sheet. The simulation results would suggest that sidewalls of suspended carbon nanotubes are more susceptible to ion damage than surfaces perpendicular to the ion beam. This is in contrast to the effects of electron irradiation on carbon nanotubes calculated by Smith and Luzzi [117]. They calculate that electrons require a kinetic energy of 86 keV before they can damage CNTs by direct knock-on, and then only atoms on surfaces normal to the beam are susceptible. However, both calculations do not take into account a supporting substrate, which influences the sputtering yield by backscattered incident ions/electrons, and also may enhance sputtering of atoms in contact with its surface by a recoil effect.

3 Scanning Tunneling Microscopy Study of OPE-A and OPE-B

In this chapter, measurements of OPE-A and OPE-B molecules on a Au(111) surface with the scanning tunneling microscope are presented. They were conducted at liquid-helium temperature in an ultra-high vacuum. Subsequently, density functional theory calculations of both molecules are compared to their observed electronic structures in Sec. 3.7.

3.1 Surface Preparation

Au(111) surfaces with single adsorbed organic molecules were prepared by dissolving the molecule under study in dichloromethane (DCM, CH_2Cl_2; also known as methylene chloride) and then allowing a drop of the solution to dry on the Au(111) surface. Beforehand, the surface was cleaned in UHV by several cycles of argon-ion sputtering and annealing at 500 °C and then shortly exposed to air in the load-lock for molecule deposition. DCM was chosen because earlier studies showed that it is possible to remove it almost completely from a Au(111) surface by gentle heating in UHV [140]. Even though DCM evaporates quickly under atmospheric conditions because of its high vapor pressure and low boiling point of 40.2 °C [141], a few molecular layers remain on the surface, which have to be desorbed by annealing the crystal in UHV. To avoid any contamination of the solvent, high-purity DCM was obtained by distilling high-pressure liquid chromatography-grade DCM twice in a clean still. The volume of the drop was $\approx 8\,\mu l$ and its evaporation typically took no longer than 20 seconds. The load-lock was then evacuated and the crystal transferred back into the UHV chamber.

The choice of molecular concentration in the drop of solvent is critical. It has to be low enough for a single drop to yield less than a monolayer of molecules on the surface. The gold crystal used had a surface area of $\approx 4 \times 6\,mm^2$ and the volume of the drop used was $\approx 8\,\mu l$. Estimating the surface area of a single molecule to $2 \times 5\,nm^2$, one can calculate the concentration for a coverage with one monolayer of OPE-A to $\approx 1\,\mu g/ml$. Although it is unlikely that the evaporation of the drop distributes the molecules evenly on the surface [142], this calculation can serve as a rough guideline.

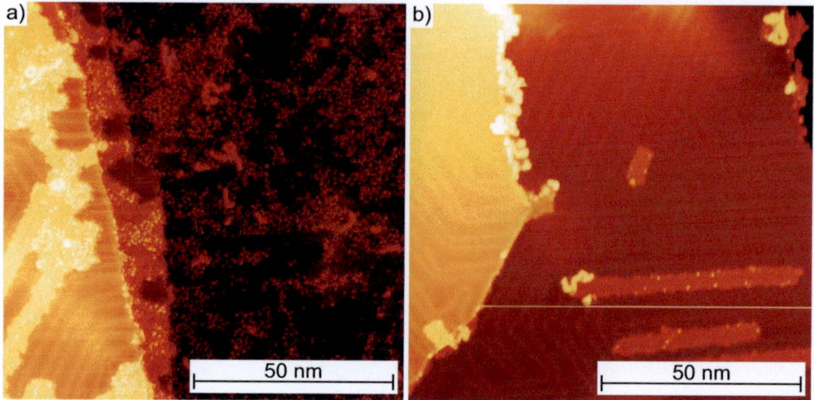

Figure 3.1: Dichloromethane on a Au(111) surface after annealing at a) 70 °C and b) 120 °C in UHV. STM parameters: a) $I = 100\,\mathrm{pA}$, $V = 1.1\,\mathrm{V}$ b) $I = 70\,\mathrm{pA}$, $V = 1\,\mathrm{V}$.

A concentration of 1 µg/ml corresponds to roughly 50 parts per *billion* of OPE-A in DCM. This underlines that even commercial HPLC grade solvents cannot be used without further purification, as they are usually specified with an evaporation residue of ≈ 5 parts per *million*. For OPE-A, two molecular concentrations were tested, with ≈ 1 µg/ml and ≈ 0.02 µg/ml, respectively. For OPE-B, only one concentration with ≈ 1 µg/ml was tested.

3.1.1 Dichloromethane on Au(111)

To be able to discern any solvent residue or contamination from the molecules under study, the Au(111) surface was first prepared with pure solvent and annealed at different temperatures to determine the temperature needed to promote the desorption of the solvent. STM images of the surface after annealing at 70 °C and 120 °C for one hour can be seen in Fig. 3.1. After annealing at 70 °C, the Au(111) reconstruction can already be seen in some areas, but most of the surface is still covered by up to two layers of residue. However, after annealing at 120 °C, most of the solvent has been desorbed. A closer look at the remaining islands is shown in Fig. 3.2. A sub-monolayer of DCM, surrounded by the Au(111) surface, is shown in Fig. 3.2a. The same appearance of parallel rows of molecules is observed through the holes of an incomplete second layer in Fig. 3.2b.

Bilayers of DCM are shown in Figs. 3.2c and 3.2d. DCM forms parallel lines of bright dots or arranges in concentric (half-)circles. Whether these structures stem from the DCM itself or from some impurity in our ultra-clean solvent is unclear. However, through these experiments it is at least possible to distinguish them from the organic

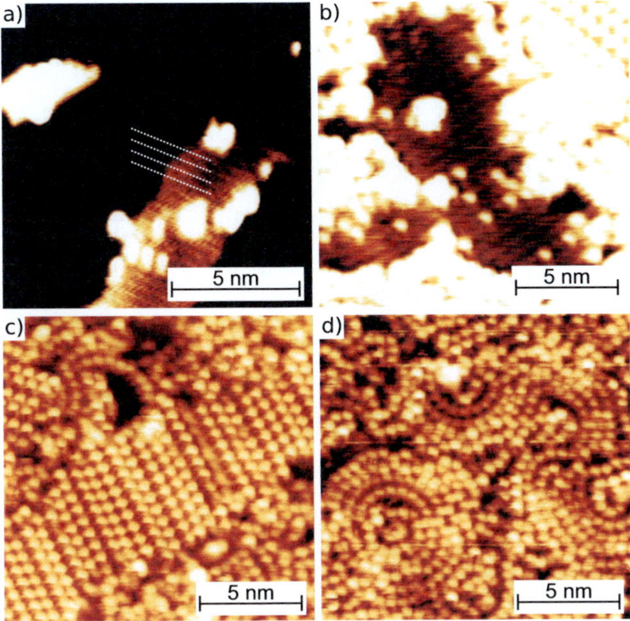

Figure 3.2: Layers of dichloromethane on Au(111). The first layer of DCM is ordered in parallel rows, seen a) directly on the gold surface and b) through the holes of another layer on top. The second layer of DCM arranges in c) parallel lines of molecules and d) concentric circles. STM parameters: a-c) $I = 100\,\text{pA}$, $V = -1\,\text{V}$ d) $I = 35\,\text{pA}$, $V = 0.8\,\text{V}$.

molecules under study. All further samples treated in this chapter were annealed at 120 °C after molecule deposition.

3.2 Self-Assembly of OPE-A and OPE-B

Both OPE-A and OPE-B arrange in regular patterns on the gold surface, see Fig. 3.3 and Fig. 3.4. They reside preferentially on the face-centered-cubic region of the Au(111) reconstruction. In both cases they arrange either side-by-side or in lines of pairs. The molecules lie flat on the surface, i.e. the benzene rings of the phenylene and phenanthrene groups are aligned parallel to the surface. The five (two plus two) hexyl chains of OPE-A (OPE-B) can be clearly seen at the sides of each molecule. This planar configuration seems to be energetically favored. The alkyl chains of neighboring molecules are seen to interdigitate in most cases.

At the single bonds, along the backbone of the molecule, the subunits and the anchor groups of a molecule can rotate against each other. As the hexyl chains at the

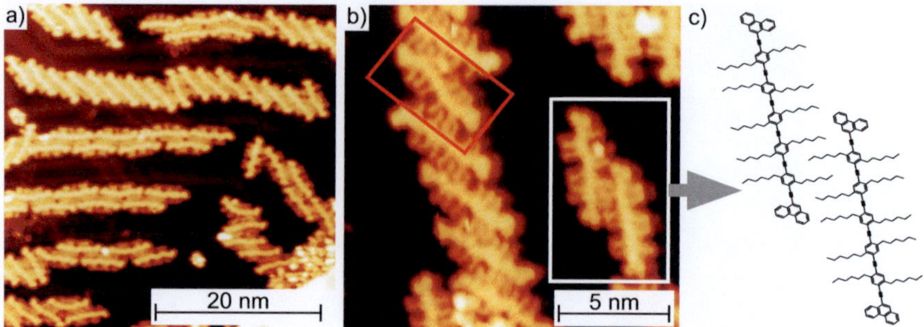

Figure 3.3: a) Arrangement of OPE-A molecules on a Au(111) terrace, on the fcc region of the Au(111) reconstruction. Molecules are positioned side-by-side or in parallel chains. b) Closer view of the assembly. Hexyl chains of neighboring molecules can be seen to interdigitate. The observed asymmetry in the position of the hexyl chains as well as the asymmetry of the anchor groups allow to infer the conformation of a molecule from the "topographic" image, as shown in c) for the molecules indicated in b). An OPE-A molecule with a strongly bent backbone is marked in red. STM parameters: a) $I = 100\,\text{pA}$, $V = 2\,\text{V}$ b) $I = 40\,\text{pA}$, $V = -2\,\text{V}$.

phenylene groups are attached asymmetrically, the observed distance between neighboring chains of an individual molecule varies, when its subunits are not in the same rotational conformation. The position of the phenanthrene anchor groups is also not symmetric under rotation. From topographic STM images, it is possible to infer the conformation of the molecule due to these two effects, see Fig. 3.3c and Fig. 3.4c.

While most OPE-A and OPE-B molecules assume a planar configuration on the surface, the molecule's backbone is often observed to bend, despite its supposedly rigid chemical structure [16]. OPE-A and OPE-B molecules with especially large bending angles in their backbone are indicated in Fig. 3.3 and Fig. 3.4 by a red rectangle.

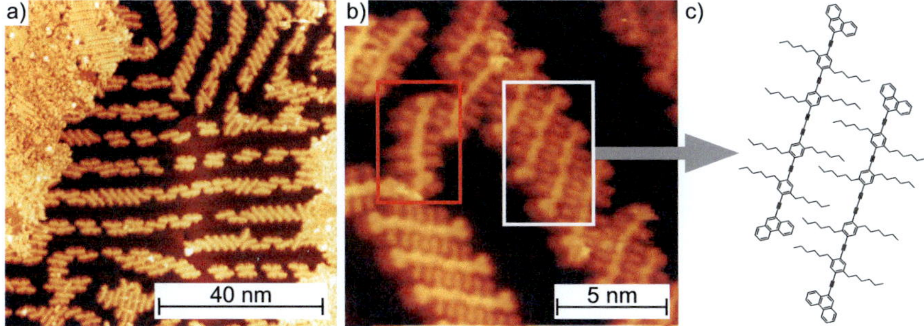

Figure 3.4: a) Arrangement of OPE-B molecules on a Au(111) terrace, again on the fcc region of the reconstruction. A similar ordering as for OPE-A molecules is observed, either side-by-side or rows-of-pairs. b) Closer view of molecules ordered side-by-side. Neighboring hexyl chains interdigitate, and in many cases it is possible to infer the conformation of a molecule from the observed "topographic" asymmetries, as shown in c) for the molecules indicated in b). An OPE-B molecule with a strongly bent backbone is marked in red. STM parameters: a) $I = 100\,\mathrm{pA}$, $V = 1.5\,\mathrm{V}$ b) $I = 100\,\mathrm{pA}$, $V = -1.5\,\mathrm{V}$.

3.3 Single OPE Molecules

For OPE-A, a very low molecular concentration of $\approx 0.02\,\mu\mathrm{g/ml}$ for the surface preparation allowed the observation of single molecules. Typically, only a few molecules are found per Au(111) terrace, with roughly half of the surface covered by large islands of solvent residue, as shown in Fig. 3.5. The molecules tend to arrange in pairs or triples even at this very low concentration, see Fig. 3.6 for examples. This suggests that the molecules are mobile at higher temperatures and diffuse on the Au(111) surface. A side-by-side arrangement of molecules appears to be energetically favored, which is why they freeze out in this configuration upon cooling.

Whilst the energy landscape of the Au(111) is flat enough for the molecules to diffuse at higher temperatures, it is not entirely structureless: As described in Sec. 2.2.2, the "elbow" sites of the reconstruction offer the lowest electron binding energies on the surface, followed by the fcc regions. Consequently, single molecules and isolated pairs are often observed at these "elbow" sites. Arrangements of OPE-A and OPE-B are typically found on the fcc regions of the surface, as described in Sec. 3.2.

After deposition of OPE-B, surface regions can be identified where the molecules did not arrange as orderly as shown in Fig. 3.4. Instead, the molecules are positioned rather freely, with some groups of molecules even lying on the transition regions of the surface reconstruction, as shown in Fig. 3.7a. In those regions, single OPE-B molecules are observed as well, see Fig. 3.7b/c. Clusters of OPE-B molecules arrange similar to OPE-A molecules: Side-by-side in such a way that the hexyl chains of neighboring molecules interdigitate, as shown in Fig. 3.8.

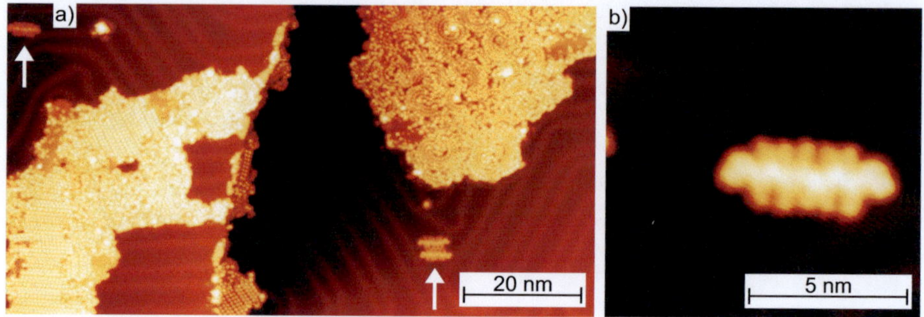

Figure 3.5: At low concentrations, typically less than one OPE molecule is found per $50 \times 50\,\text{nm}^2$ area, isolated from large blotches of solvent residue. STM parameters: $I = 100\,\text{pA}$, $V = -2\,\text{V}$.

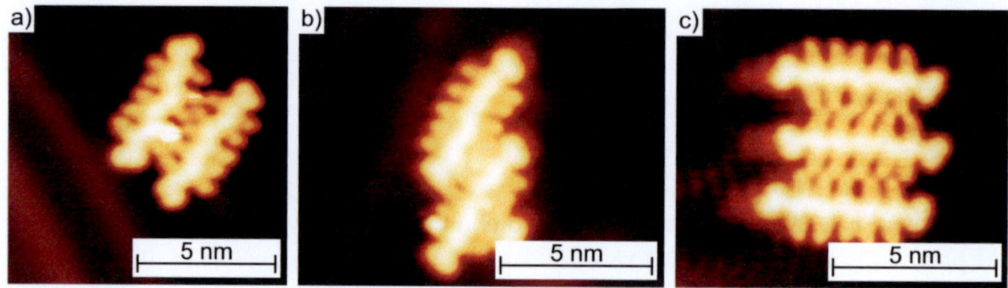

Figure 3.6: Pairs/triple of OPE-A molecules. The alkyl chains of neighboring molecules are interdigitating in some cases. STM parameters: $I = 125\,\text{pA}$, $V = -1.5\,\text{V}$ b) $I = 50\,\text{pA}$, $V = 1\,\text{V}$ c) $I = 100\,\text{pA}$, $V = -300\,\text{mV}$.

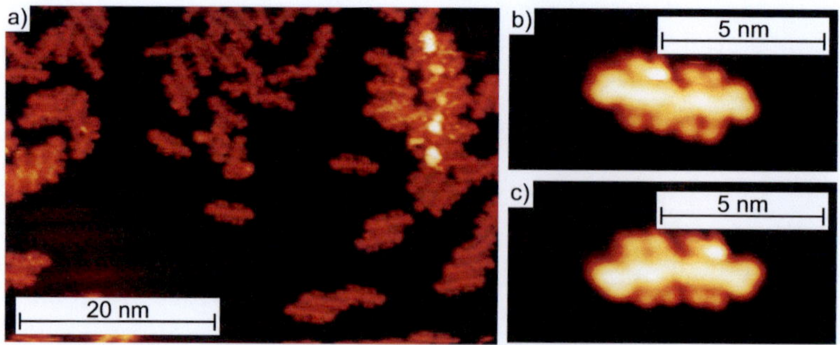

Figure 3.7: a) A region on the Au(111) surface where the OPE-B molecules are not arranged in an ordered manner. b),c) Closer views of two single molecules found in that region. STM parameters: $I = 100\,\text{pA}$, $V = -1.5\,\text{V}$.

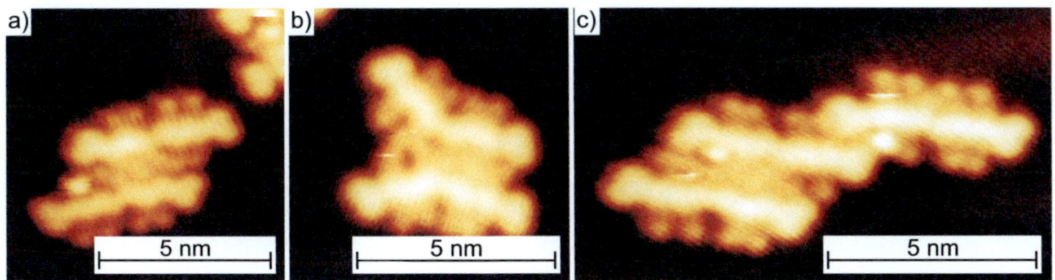

Figure 3.8: Pairs/triple of OPE-B molecules. STM parameters: a) $I = 100\,\text{pA}$, $V = -1.5\,\text{V}$ b) $I = 400\,\text{pA}$, $V = 1\,\text{V}$ c) $I = 100\,\text{pA}$, $V = -1.5\,\text{V}$.

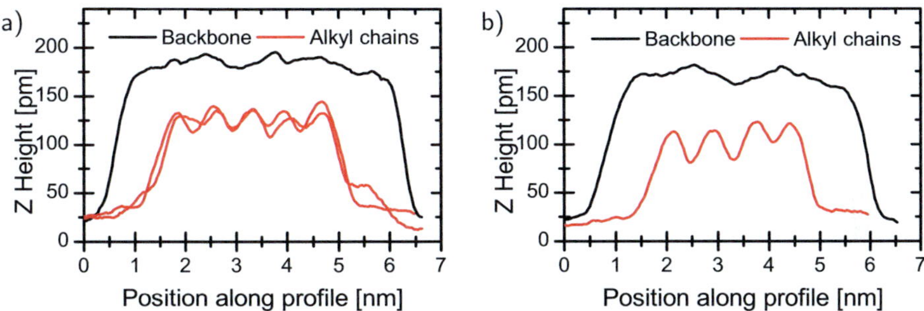

Figure 3.9: Height profiles of individual a) OPE-A and b) OPE-B molecules, recorded at a bias voltage of -1.5 V and with a current of 100 pA along the backbone of each molecule and parallel to the backbone on the hexyl chains. The backbone appears higher than the alkyl chains. In agreement with their chemical structure, OPE-A molecules are longer than OPE-B molecules.

Height profiles over single OPE-A and OPE-B molecules are plotted in Fig. 3.9. They were recorded at a bias voltage of -1.5 V and with a current of 100 pA along the backbone of each molecule and parallel to the backbone on the hexyl chains. The apparent height of both molecules is approximately 175 pm, with the backbone appearing higher than the hexyl chains. The OPE-A molecule has a length over its backbone of $\approx 6\,\text{nm}$, whereas a length of $\approx 5.5\,\text{nm}$ is recorded for OPE-B, in good agreement with the shorter structure of OPE-B.

3.3.1 Moving Molecules on the Surface

By positioning the STM tip over the anchor group of a molecule and increasing the current to 2 nA, it was possible to drag OPE-A by the STM tip, see Fig. 3.10. When the current was decreased to normal imaging levels again, the molecule was fixed at its new position. With a suitable tip, molecules could be freely moved on a Au(111) terrace, even across transition regions of the surface reconstruction. As shown in the

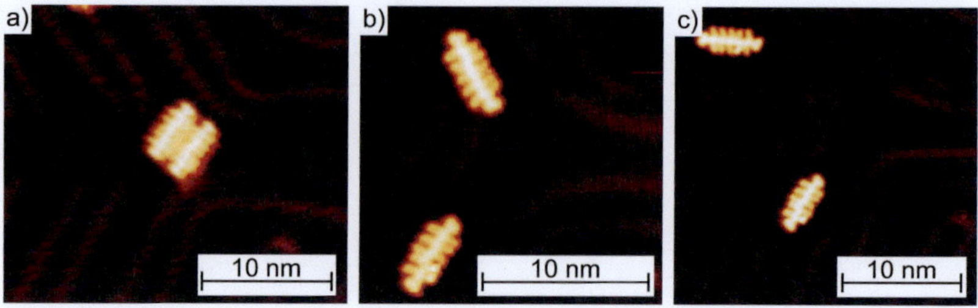

Figure 3.10: Pulling an OPE-A molecule. The tip is placed over the anchor group and the current increased to 2 nA. When the tip is moved, the molecule is following in the wake of the tip's electric field. Images were taken after pulling. STM parameters: a) $I = 50\,\text{pA}$, $V = 1.5\,\text{V}$ b) $I = 40\,\text{pA}$, $V = 1.5\,\text{V}$ c) $I = 50\,\text{pA}$, $V = 1.5\,\text{V}$.

consecutive images in Fig. 3.10, it was possible to pull apart a pair of molecules, isolating them. We suspect this pulling effect is driven by the high electric field gradients generated by a sharp STM tip. The molecules did not follow the position of STM tips that appeared to be blunt, even at higher tunneling current setpoints of up to 100 nA.

By another technique, described by Jung et al. [143] it proved possible to move molecules on the surface independent of the shape of the STM tip: The feedback loop was disabled and the STM tip moved closer to the surface in the vicinity of a molecule. Then, the tip was moved laterally over the molecule. Eventually, repulsive forces between the tip and the molecule pushed it away from the tip, more or less in the direction of tip movement. Then, the feedback loop was re-activated and a topographic image recorded. Using this technique, an OPE-A molecule could be de-attached from an island of solvent residue and moved further around on the surface, see Fig. 3.11 for the sequence of recorded images after individual pushing actions. Both possibilities of moving single molecules on the Au(111) surface indicate that they lie on a comparably flat potential energy landscape.

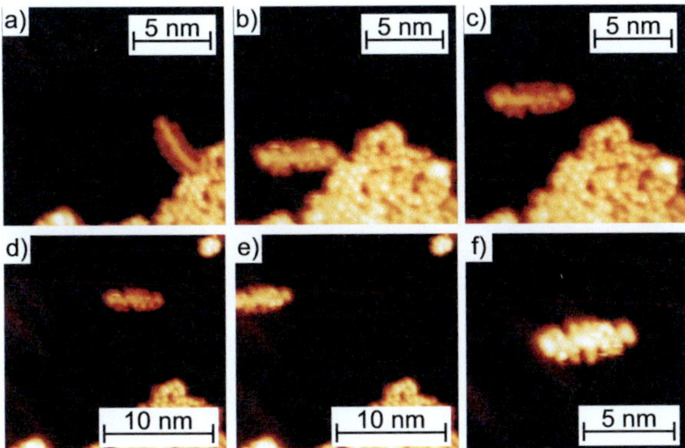

Figure 3.11: Pushing an OPE-A molecule by repulsive forces. The tip is placed close to the gold surface near the molecule and the feedback loop disabled. Moving the tip over the molecule pushes it around on the surface. The feedback loop is then restored for imaging. STM parameters: a) $I = 30\,\mathrm{pA}$, $V = 1.5\,\mathrm{V}$ b) $I = 100\,\mathrm{pA}$, $V = 1.5\,\mathrm{V}$ c) $I = 100\,\mathrm{pA}$, $V = 1.5\,\mathrm{V}$ d) $I = 60\,\mathrm{pA}$, $V = 1.5\,\mathrm{V}$ e) $I = 85\,\mathrm{pA}$, $V = 1.5\,\mathrm{V}$ f) $I = 150\,\mathrm{pA}$, $V = 1.5\,\mathrm{V}$.

3.3.2 Contacting with the STM tip

Efforts were undertaken to contact single molecules with the STM tip for direct I-V measurements. At low bias voltages of 200 mV and 300 mV, the STM tip was placed above one of the anchor groups of a molecule and the feedback loop disabled. The tip was then slowly lowered towards the surface while monitoring the tunneling current. Ideally, if there is an attraction between the molecule and the STM tip, the molecule attaches to the latter upon approach and can be pulled upwards from the surface upon retraction. The measured current during retraction typically exceeds that of mere tunneling, as long as the molecule is acting as a conductive bridge between tip and surface.

Such approach-retract curves over the anchor group of an OPE-A molecule at bias voltages of 200 mV and 300 mV are plotted in Fig. 3.12 and Fig. 3.13. With the feedback loop active, the tunneling current was stabilized at 100 pA. The corresponding height defines the zero point of the height scale of the approach-retract curves. The feedback loop was then disabled and the tip first retracted by 100 pm, before it was lowered until a current of $\approx 400\,\mathrm{nA}$ was recorded. Then, the tip was retracted again. The current was continuously monitored during this process.

For a bias of 200 mV, the current increased exponentially during approach up to a value of $\approx 150\,\mathrm{nA}$. Then, its slope changed. This indicated that the STM tip had come into contact with the anchor group. Upon retraction of the STM tip, the current stayed at a high level, with two steps observed in the current. Eventually, the current fell back to the tunneling level upon disconnection of the molecule from the STM tip when the feedback loop was reactivated (not shown, as the data was not recorded).

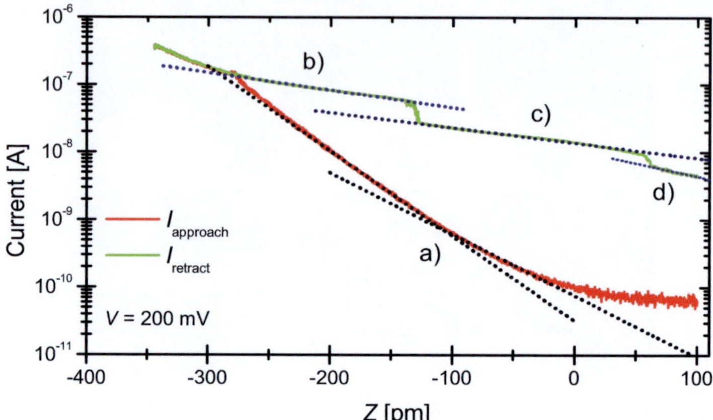

Figure 3.12: Contacting an OPE-A molecule by pressing the STM tip onto its anchor group. The slope of the current changes on approach, and the current stays at a high level upon retraction of the STM tip, indicating an attachment of the anchor group to the tip. Eventually, the molecule disconnects from the tip (not shown) and the current returns to the level of tunneling. During retraction, steps are observed in the current.

A similar approach-retract curve was recorded at a bias voltage of 300 mV, except for earlier disconnection of the molecule from the STM tip at a height around the zero level. Typically, higher bias voltages inhibited an attachment of the molecule to the STM tip.

The data shall first be analyzed in the terms of a tunneling system, to demonstrate that this analysis does not produce physically meaningful results for the current upon withdrawal of the STM tip. Using the WKB approximation of the tunneling matrix element from Sec. 2.2 the tunneling current for small bias voltages can be approximated with:

$$I \propto \exp\left(-\frac{2Z\sqrt{2m_\mathrm{e}}}{\hbar}\sqrt{\bar{\phi}+eV}\right) \tag{3.1}$$

Here m_e denotes the electron mass, Z the tip-sample distance and $\bar{\phi}$ the mean work function of the electrodes. Exponential fits to the approach-retract curves are plotted in Fig. 3.12 and Fig. 3.13 for all regions of interest. From the fits mean work functions during approach between 3.95 eV and 7.75 eV and between 3.56 eV and 6.97 eV are obtained for bias voltages of 200 mV and 300 mV, respectively, depending on the selected fitting region. These values are reasonable for tunneling between gold electrodes, as the work function of a gold surface is ≈ 5.3 eV [144]. However, for the current upon retraction, the exponential fits yield unphysical values for the mean work function: 48 meV, 138 meV and 490 meV for a bias voltage of 200 mV and 376 meV and 192 meV

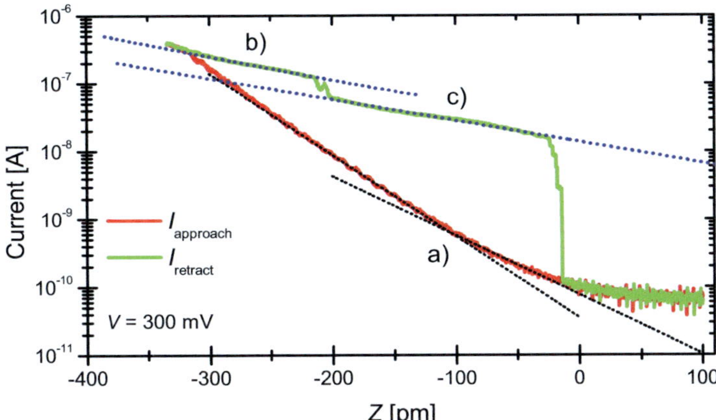

Figure 3.13: Contacting an OPE-A molecule by pressing the STM tip onto its anchor group. For both approach-retract curves, the slope of the current changes on approach, and the current stays at a high level upon retraction of the STM tip, indicating an attachment of the anchor group to the tip. Eventually, the molecule disconnects from the tip and the current returns to the level of tunneling. During retraction, steps are observed in the current.

for a bias voltage of 300 mV. These are of the same magnitude or even smaller than the applied bias voltage, which is not possible.

Instead of influencing the work function of the tunneling system, the molecule acts as a conductive bridge between the STM tip and the gold surface. The molecule is now considered as a chain of individual rigid elements that can bend relative to each other. At first, only the first part of this chain (i.e., the anchor group of the molecule) is lifted by the STM tip, until the strain causes a certain critical stress and bending angle on the anchor group's bond. Then, the second element of the chain is detached from the surface. For an illustration of this toy model, see Fig. 3.14. We identify the steps in the current signal during retraction with such detachment events. At each step, another part of the molecule is detached from the surface. As the length of an OPE-A molecule is ≈ 50 Å, it is likely that only a small part of the molecule was lifted from the surface.

The conductance $G_{\text{retr.-appr.}} = (I_{\text{retract}} - I_{\text{approach}})/V$ for both approach-retract curves is plotted in Fig. 3.15 and Fig. 3.16 in units of the conductance quantum $G_0 = \frac{2e^2}{h} = 7.748 \cdot 10^{-5}$ S. For a bias voltage of 200 mV (300 mV), three (two) regions of the curve, separated by steps in the conductance, were fitted linearly. The current upon approach is subtracted from the current upon withdrawal to yield only the current flowing through the molecule. Coherent quantum transport through a many-channel junction can be described in the Landauer-Büttiker formalism [145]. The total conductance G is the sum over N conductance channels, each with a conductance of G_0 and a transmission coefficient T_n:

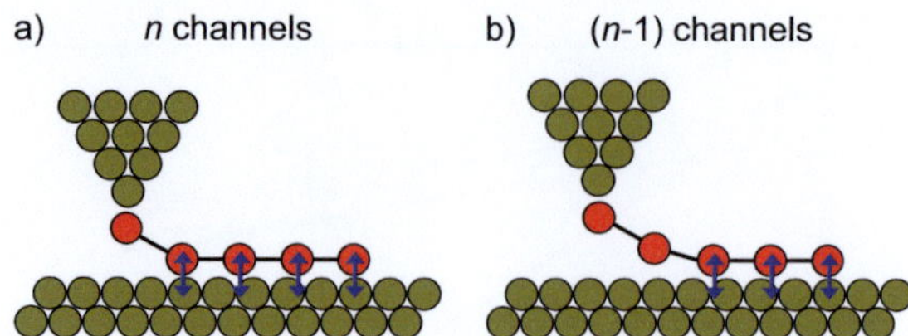

Figure 3.14: The molecule is considered as a chain of rigid segments, bending relative to each other. a) At first, only the first part of the chain is lifted, until a critical bond angle/elongation is reached. b) Then, the second part of the chain is detached from the surface to mitigate the strain. The number of conductance channels from the molecule to the substrate is reduced.

$$G = \frac{2e^2}{h} \sum_{n=1}^{N} T_n \tag{3.2}$$

By lifting parts of the molecule from the surface, the number of parallel conductance channels from the molecule to the substrate is reduced. For a bias voltage of 200 mV, the conductance decreases at the steps from $3.69 \cdot 10^{-3}\, G_0$ to $1.67 \cdot 10^{-3}\, G_0$ and from $0.63 \cdot 10^{-3}\, G_0$ to $0.39 \cdot 10^{-3}\, G_0$, respectively. For a bias voltage of 300 mV, only one step with a conductance decrease from $4.78 \cdot 10^{-3}\, G_0$ to $2.19 \cdot 10^{-3}\, G_0$ was recorded. Between the steps in conductance, a linear dependence of the conductance on the withdrawal distance was observed. Possibly, a bend is introduced into the backbone of the molecule, gradually diminishing the transmission coefficient along the conductance channels of the molecular backbone. The recorded slopes are quite similar for 200 mV and 300 mV bias, ranging from $-1.20 \cdot 10^3\, \frac{\mathrm{S}}{\mathrm{m}}$ to $-2.29 \cdot 10^2\, \frac{\mathrm{S}}{\mathrm{m}}$ and from $-2.15 \cdot 10^3\, \frac{\mathrm{S}}{\mathrm{m}}$ to $-6.42 \cdot 10^2\, \frac{\mathrm{S}}{\mathrm{m}}$. The similar steps in conductance and the slopes suggest that in both lifting attempts the same parts of the molecule were detached from the surface. However, the conductance values at the steps seem not be multiples of a single value; more data is needed to make that assessment.

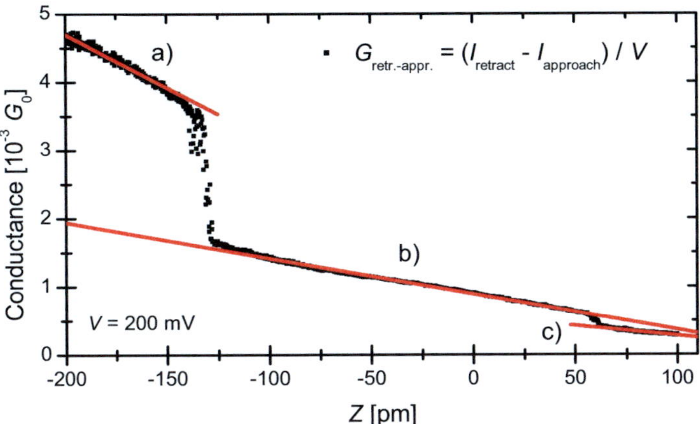

Figure 3.15: Observed conductance during contact with the molecule in units of the conductance quantum. Three regions of the curve, separated by steps in the conductance, were fitted linearly. Slopes: a) $-1.20 \cdot 10^3 \frac{\mathrm{S}}{\mathrm{m}}$ b) $-4.06 \cdot 10^2 \frac{\mathrm{S}}{\mathrm{m}}$ c) $-2.29 \cdot 10^2 \frac{\mathrm{S}}{\mathrm{m}}$.

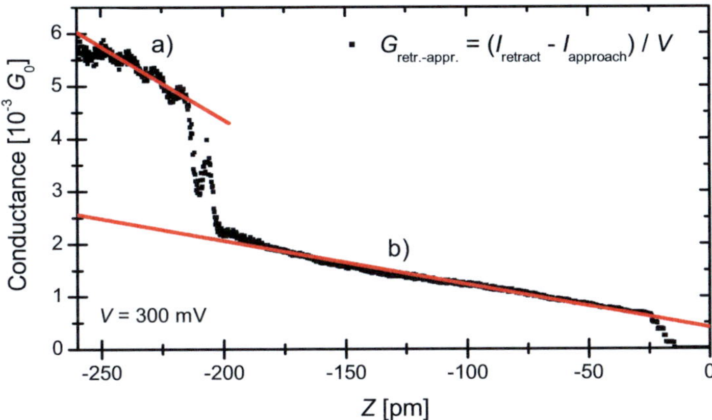

Figure 3.16: Observed conductance during contact with the molecule in units of the conductance quantum. Two regions of the curve, separated by steps in the conductance, were fitted linearly. Slopes: a) $-2.15 \cdot 10^3 \frac{\mathrm{S}}{\mathrm{m}}$ b) $-6.42 \cdot 10^2 \frac{\mathrm{S}}{\mathrm{m}}$.

3.4 Scanning Tunneling Spectroscopy

Spectroscopic measurements were recorded by positioning the STM tip over various parts of the molecule and disabling the feedback loop. Then, the bias voltage was swept from e.g. -3 V to 3 V and the current as well as the $\mathrm{d}I/\mathrm{d}V$ signal were recorded, using a lock-in amplifier. Positive bias voltages probe unoccupied states of the sample, whereas negative voltages probe occupied states. Before any measurements, spectra on the clean gold surface were acquired, such as those shown in Fig. 3.17. If the Au(111) surface state at ≈-470 mV was not observed or other instabilities were recorded, the STM tip was re-shaped by voltage pulses or by gently crashing it into the surface. Except for the surface state, no features should be discernible on spectra of clean gold. E.g., in the spectra in Fig. 3.17b a peak at ≈0.75 V is observed, which is consequently assumed to be caused by an increased density of states of the STM tip, and not attributed to molecular states if it is found in spectra over a molecule.

The various measurement positions over OPE-A and OPE-B molecules are indicated in Fig. 3.18. For OPE-A, measurements over the anchor group and close its side, as well as over the backbone and over the hexyl chains were performed. Over OPE-B molecules, a statistically significant number of spectra could only be recorded over the backbone and over the anchor groups. For each of the measurement positions above the molecules several spectra were recorded and averaged.

The $\mathrm{d}I/\mathrm{d}V$ signal over varios positions of OPE-A molecules are plotted in Fig. 3.19. The normalized $(\mathrm{d}I/\mathrm{d}V)/(I/V)$ signal is plotted in Fig. 3.20. Several features can be identified in the spectra. For negative voltages, apart from the gold surface state, several peaks can be identified: Spectra over the anchor group display a pronounced peak at ≈-1.6 V, whereas spectra close to its sides show a peak at ≈-2.3 V. The local density of states on the backbone as well as on the alkyl chains increases for negative voltages as well, but no additional peak is observed, which is not already seen in spectra of the gold surface. Spectra over the OPE-A molecule show a peak at ≈2.8 V. From the data, we can extract a HOMO-LUMO gap for OPE-A of ≈4.4 eV, as indicated in Fig. 3.20.

For the OPE-B molecule, spectra on the anchor group, the backbone and on the gold surface are plotted in Fig. 3.21, with the $(\mathrm{d}I/\mathrm{d}V)/(I/V)$ signal plotted in Fig. 3.22. Over the anchor groups, several peaks can be identified for negative voltages at ≈-1.2 V, ≈-1.4 V, ≈-2 V and ≈-2.3 V. Over the backbone, a small peak in the density of states is observed at ≈-1.2 V as well, with the density of states subsequently decreasing, then increasing to another peak at ≈-2.2 V. For positive voltages up to 1.5 V, several ripples are observed in all spectra, including those taken on the gold surface. Hence these peaks are attributed to states on the STM tip. At ≈2 V, a peak is observed in the spectra over the molecule, which is not present in spectra over the gold surface. A HOMO-LUMO gap of ≈3.2 eV can be extracted if the latter peak is attributed to the lowest unoccupied molecular orbital of OPE-B.

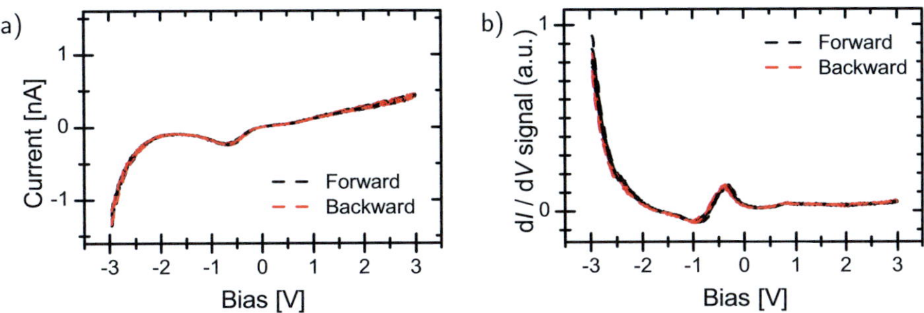

Figure 3.17: STM spectra acquired on a clean gold surface. The dI/dV signal was recorded simultaneously with a lock-in amplifier. The density of states is flat for positive voltages and eventually increases towards large negative voltages. The surface state of gold is clearly visible.

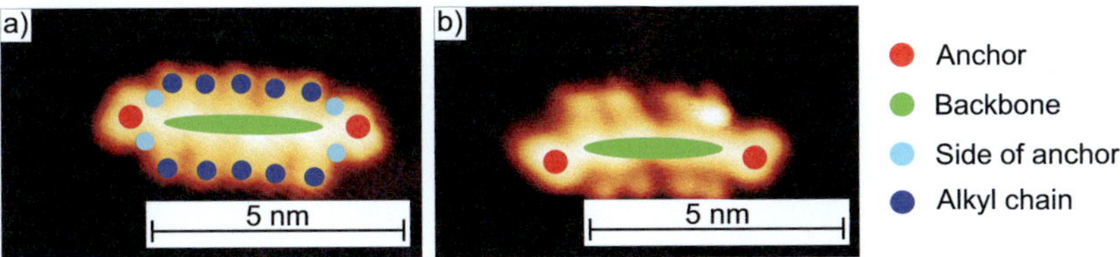

Figure 3.18: The positions over a) OPE-A and b) OPE-B where spectroscopic measurements were acquired.

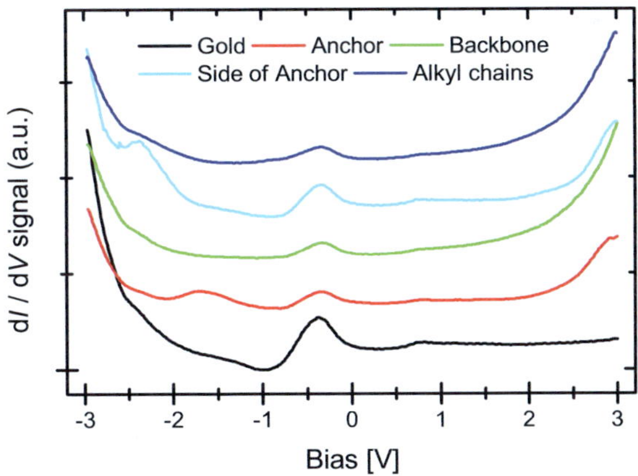

Figure 3.19: Spectroscopy on OPE-A. Averaged dI/dV signal over various positions of the molecule.

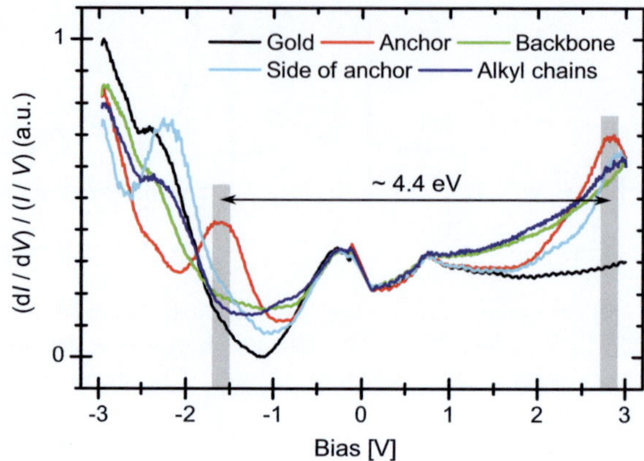

Figure 3.20: Spectroscopy on OPE-A. Averaged $(\mathrm{d}I/\mathrm{d}V)/(I/V)$ signal over various positions of the molecule. The Au surface state and the peak at $\approx 0.75\,\mathrm{V}$ are not attributed to the molecule. The energy difference between the highest occupied and the lowest unoccupied state (HOMO-LUMO gap) then is $\approx 4.4\,\mathrm{eV}$.

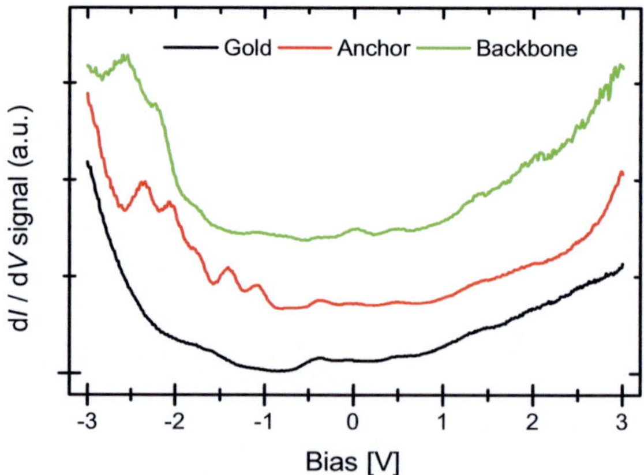

Figure 3.21: Spectroscopy on OPE-B. Averaged $\mathrm{d}I/\mathrm{d}V$ signal over various positions of the molecule.

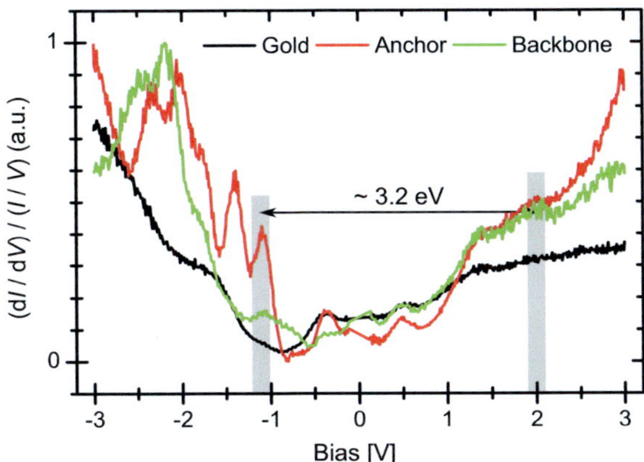

Figure 3.22: Spectroscopy on OPE-B. Averaged $(\mathrm{d}I/\mathrm{d}V)/(I/V)$ signal over various positions of the molecule. Peaks between -1 V and 1.5 V are not attributed to the molecule, as they are also evident in spectra on the Au surface. The energy difference between the highest occupied and the lowest unoccupied state (HOMO-LUMO gap) then is $\approx 3.2\,\mathrm{eV}$.

As different STM tips and lock-in-amplifier settings were used for the measurements over OPE-A and OPE-B, the absolute values of the $(\mathrm{d}I/\mathrm{d}V)/(I/V)$ signal are not comparable. For both molecules more pronounced peaks in the density of states are observed for negative voltages than for positive voltages. Band gaps of $\approx 4.4\,\mathrm{eV}$ and $\approx 3.2\,\mathrm{eV}$ can be extracted for OPE-A and OPE-B, respectively. From ultraviolet-visible fluorescence measurements on solutions of OPE-A and OPE-B in dichloromethane, we extract band gaps of 2.95 eV for OPE-A and 2.90 eV for OPE-B, see Sec. 2.1.1. Due to a solvatochromic shift from the polarizability of the surrounding solvent, these values need not correspond to the band gaps of the free molecules. Likewise, STM measurements are influenced by hybridization of the molecular orbitals with surface states. For both molecules, the energy level of the LUMO is difficult to ascertain from the STM spectra. For OPE-A, only one peak is observed for positive voltages that is not attributed to the surface or the STM tip, and for OPE-B only weak signs of a peak are found for positive bias voltages. In spite of this, a good agreement between STM and fluorescence data for OPE-B is found. For an OPE of similar length, a band gap of $\approx 3\,\mathrm{eV}$ was observed by ultraviolet-visible spectrophotometry by Lu et al[25], again in good agreement. In light of this, the band gap of OPE-A of $\approx 4.4\,\mathrm{eV}$ from STM spectra is likely overestimated. Later, these spectroscopic results will be compared to density functional theory calculations in Sec. 3.7.

3.4.1 Conductance Maps

To be able to assign individual peaks of the spectra to orbitals of the molecules, conductance maps were recorded: At various bias voltages, $\mathrm{d}I/\mathrm{d}V$ images were obtained of both molecules using a lock-in amplifier. At a constant tip-sample separation, these conductance maps correspond to images of the local density of states (LDOS), i.e., of the molecular orbitals. In constant-current mode, the conductance data contains a convolution of the LDOS at different heights with the tunneling conductance, which has an exponential height dependence. To keep the tunneling current constant when the tip moves to a region of lower LDOS, the feedback loop decreases the tip-sample distance, thereby increasing the tunneling conductance and hence the $\mathrm{d}I/\mathrm{d}V$ signal. However, that increase in $\mathrm{d}I/\mathrm{d}V$ signal does not exactly cancel out the decrease in $\mathrm{d}I/\mathrm{d}V$ signal due to the lower LDOS. Consequently, the conductance maps obtained in constant-current mode still correspond to a map of the local density of states of the sample.

Conductance maps of the OPE-A molecule at various positive and negative bias voltages are shown in Fig. 3.23. For negative bias voltages, two different orbitals are observed:

- At -2.5 V (see Fig. 3.23a) two lobes on each side of a phenylene-ethynylene subunit are observed. We will call this orbital HOMO-1.
- At bias voltages of -2 V (not shown), -1.5 V (see Fig. 3.23b) and -1 V (not shown) an orbital with three lobes per phenylene-ethynylene subunit is observed. The elliptical lobes are centered along the backbone of the molecule. We identify this orbital as the highest occupied molecular orbital (HOMO), in agreement with STM spectroscopy, where the highest unoccupied orbital was observed at $\approx$-1.6 V.

At bias voltages between -1 V and 1.5 V (see Fig. 3.23c), the molecule is hardly discernible in the $\mathrm{d}I/\mathrm{d}V$ image, indicating that these voltages lie in the energy gap of OPE-A. At a bias of 2 V (see Fig. 3.23d), it is still difficult to make out the orbital structure, even though the structure of the molecule is again discernible. From STM spectroscopy, the lowest unoccupied orbital of OPE-A is expected at a bias voltage of $\approx$2.8 V. However, sufficiently stable imaging conditions could not be established at bias voltages greater than 2 V. As explained in Sec. 2.2.1, due to hybridization of molecular orbitals with the surface states a clear image of the LUMO may not be attainable in any case. In the case of OPE-A, the hybridization appears to be stronger for the LUMO than for the HOMO.

Conductance maps of OPE-B were recorded in constant-height mode. For bias voltages of -1.5 V and -1 V (see Fig. 3.24a,b), three lobes along the backbone per phenylene-ethynylene subunit are observed. The shape of this orbital is similar to the HOMO of OPE-A and likely constitutes the HOMO of OPE-B. At positive bias voltages of 1.5 V and 2 V (see Fig. 3.24c,d) it was again not possible to obtain a clear

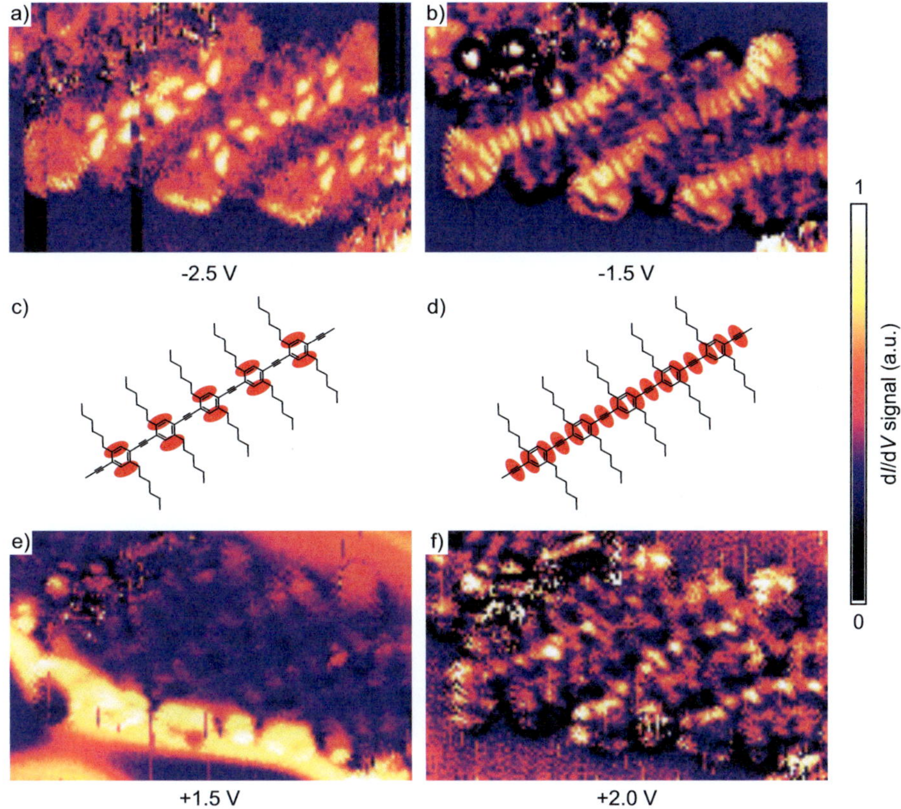

Figure 3.23: a,b,e,f) $\mathrm{d}I/\mathrm{d}V$ signal from an arrangement of OPE-A molecules recorded at different bias voltages. a) For -2.5 V, two lobes next to each phenylene-ethynylene subunit are observed, as illustrated in c). b) For -1.5 V a sub-structure of three orbitals per subunit is observed, as illustrated in d). e) At a bias voltage of 1.5 V, the structure of the molecule is only barely visible. The contrast of this image is enhanced compared to the other three. f) At a bias of 2 V, a clear orbital structure is again not observed. STM parameters: Frame size $8 \times 5\,\mathrm{nm}^2$, $I = 200\,\mathrm{pA}$, $f_{\mathrm{osc}} = 1.33\,\mathrm{kHz}$, $A_{\mathrm{osc}} = 42\,\mathrm{mV}$.

conductance map. With a different STM tip and again in constant-height mode, an image of the HOMO of an OPE-B molecule at -1.5 V was measured where the lobes of the molecular orbital show a different shape, depending on whether they lie on a phenylene or on an ethynylene group, see Fig. 3.24e. The orbital lobes on the sides of the phenylene groups thereby appear more elliptical than those on ethynylene groups, as indicated in the figure. From the spectroscopic measurements on OPE-B, the position of the HOMO was estimated to be at ≈-1.2 V, with another peak in the density of states at ≈-1.4 V. There are no discernible differences in conductance maps of the molecule at these two bias voltages. For the LUMO, an energy position at ≈ 2 V is seen in the spectra. Again, it was not possible to obtain a clear conductance map of it. We attempted to image the HOMO-1 as for OPE-A, but were not able to establish sufficiently stable imaging conditions at bias voltages below -2 V either. The shapes of the molecular orbitals of OPE-A and OPE-B presented in this chapter will be compared to the results of density functional theory calculations in Sec. 3.7.

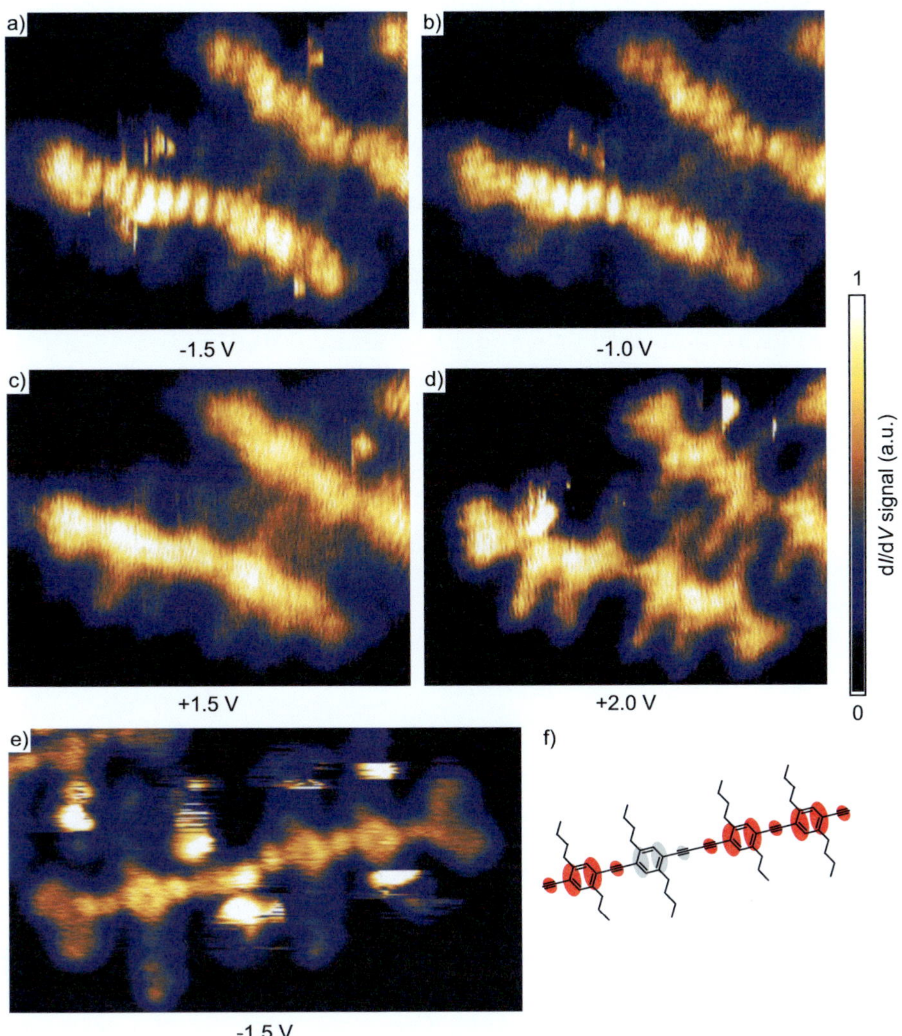

Figure 3.24: a-e) dI/dV signal from a pair of OPE-B molecules recorded at different bias voltages in constant-height mode. a,b) For negative bias voltages, the HOMO is observed. Its shape is similar to that of OPE-A and comprises three lobes per phenylene-ethynylene subunit. c,d) The LUMO could not be clearly imaged at bias voltages of 1.5 V and 2 V. e) Using a different STM tip, a shape of the HOMO was observed in the dI/dV image, where the orbital lobes showed a different elliptical shape depending on whether they were on a phenylene or an ethynylene group, as illustrated in f). STM parameters: a-d) Image size $5 \times 4\,\mathrm{nm}^2$, *Constant-height mode*, $f_{\mathrm{osc}} = 8.16\,\mathrm{kHz}$, $A_{\mathrm{osc}} = 100\,\mathrm{mV}$. e) Image size $5 \times 2.5\,\mathrm{nm}^2$, *Constant-height mode*, $f_{\mathrm{osc}} = 8.16\,\mathrm{kHz}$, $A_{\mathrm{osc}} = 100\,\mathrm{mV}$.

3.5 Atomic Resolution Imaging

As mentioned in Sec. 2.2.1, atomic resolution, i.e., a direct image of a molecule's chemical structure, is typically not achievable with an STM, as typically the imaging contrast is based on the local density of electronic states, which need not be correlated with the chemical structure. However, by using bias voltages in the energy gap it was possible to observe the atomic structure of both OPE-A and OPE-B with very high resolution.

In constant-current mode, at a set point of 90 pA and bias voltages between 100 mV and 200 mV, the individual benzene rings of OPE-A can be identified in the $\mathrm{d}I/\mathrm{d}V$ image, see Fig. 3.25. An amplitude modulation between 60 mV and 85 mV at 205 Hz was used for lock-in detection of the $\mathrm{d}I/\mathrm{d}V$ signal. The benzene rings along the backbone of the molecule are discernible as well as the three fused benzene rings of the anchor group. For comparison, the chemical structure of OPE-A was overlaid on the topographic image by scaling it, see Fig. 3.25d. The topographic image recorded simultaneously and shown in Fig. 3.25a does not show these features.

Images of even higher resolution of OPE-B could be recorded in constant-height mode, again at low bias voltages. In both the current- and the $\mathrm{d}I/\mathrm{d}V$ image the atomic structure of OPE-B is observed, see Fig. 3.26. Tunneling current images at various tip-sample distances were recorded and are shown in Fig. 3.27. The individual hexagons of the carbon rings on the backbone and in the anchor groups can be seen. As observed before for single OPE molecules, the backbone of the molecule is bent.

The tunneling current is highest over the center of a benzene ring, and decreases sharply over assumed positions of carbon atoms and covalent bonds between them. For the ethynylene groups along the backbone of the molecule, the highest current is observed on a cone-shaped area protruding from each side of the group. The observed contrast is similar to what Temirov et al. reported on PTCDA and tetracene molecules and termed scanning tunneling hydrogen microscopy [57]. To date, there are no reports of such an STM imaging contrast on other molecules. The exact cause of the contrast is still under discussion, whether it stems from Pauli repulsion [58] or a favorable hybridization of an hydrogen atom at the STM tip apex with the organic molecule under study [60]; the latter having only been theoretically discussed for PTCDA molecules. In light of this, the atomic-resolution images of large OPE molecules shown here may fuel further theoretical studies of the involved imaging contrast. Such high-resolution STM images of OPE molecules provide a fascinating insight into the structure of an Ångström-sized object at the picometer scale.

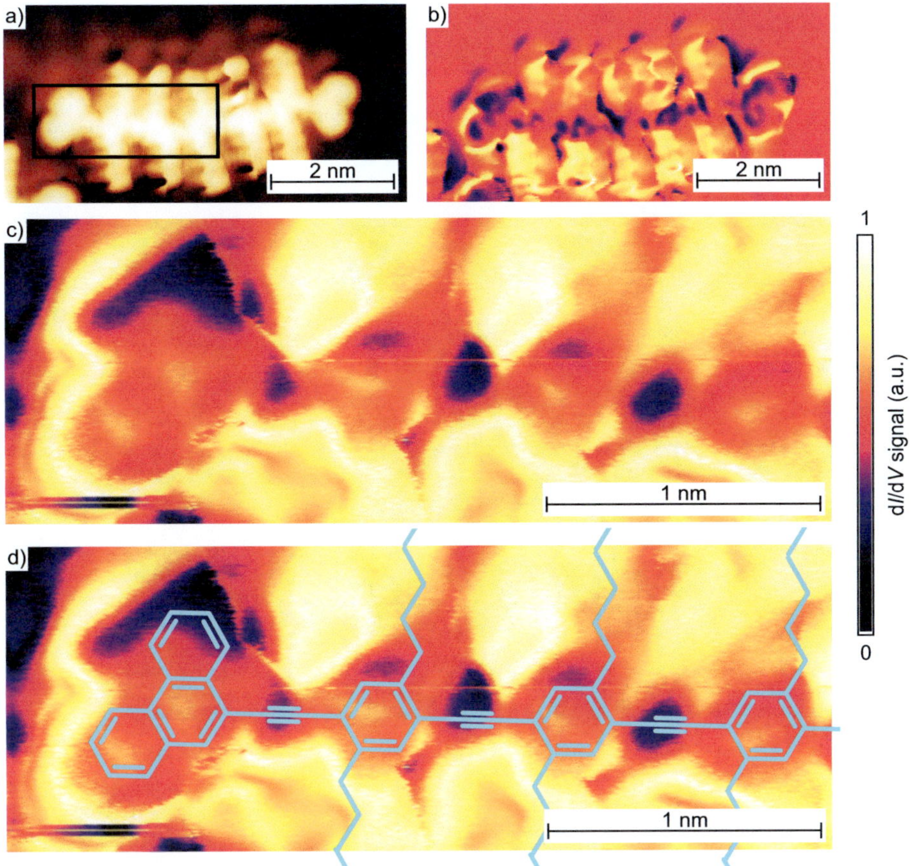

Figure 3.25: a) "Topographic" and b-d) dI/dV images of OPE-A at low bias voltages in constant-current mode. The chemical structure of the molecule was overlaid in d). STM parameters: a,b) $I = 100\,\text{pA}$, $V = 200\,\text{mV}$ $f_{\text{osc}} = 205\,\text{Hz}$, $A_{\text{osc}} = 60\,\text{mV}$. c,d) $I = 90\,\text{pA}$, $V = 100\,\text{mV}$ $f_{\text{osc}} = 205\,\text{Hz}$, $A_{\text{osc}} = 85\,\text{mV}$.

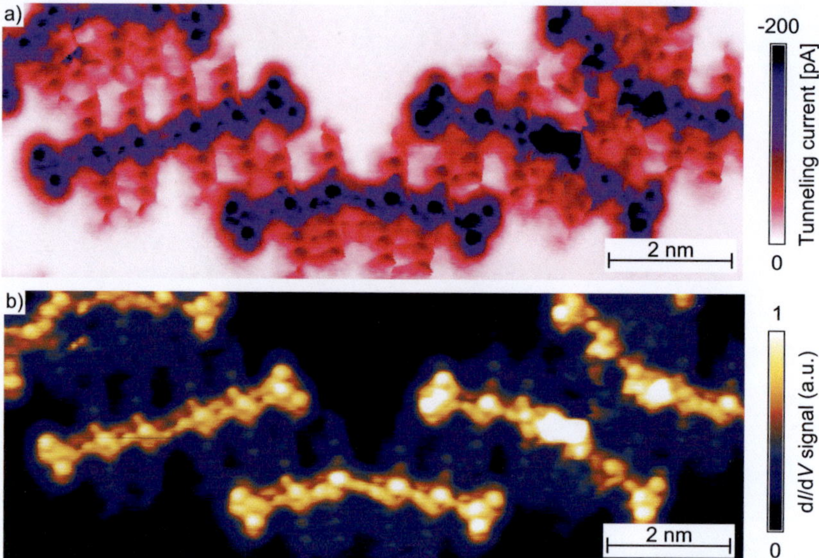

Figure 3.26: a) Tunneling current and b) dI/dV images of OPE-B at low bias voltages in constant-height mode. Individual benzene rings of the anchor groups as well as of the phenylene groups are visible in both images. STM parameters: *Constant-height mode*, $V = -7.5\,\mathrm{mV}$, $f_{\mathrm{osc}} = 15.65\,\mathrm{kHz}$, $A_{\mathrm{osc}} = 100\,\mathrm{mV}$.

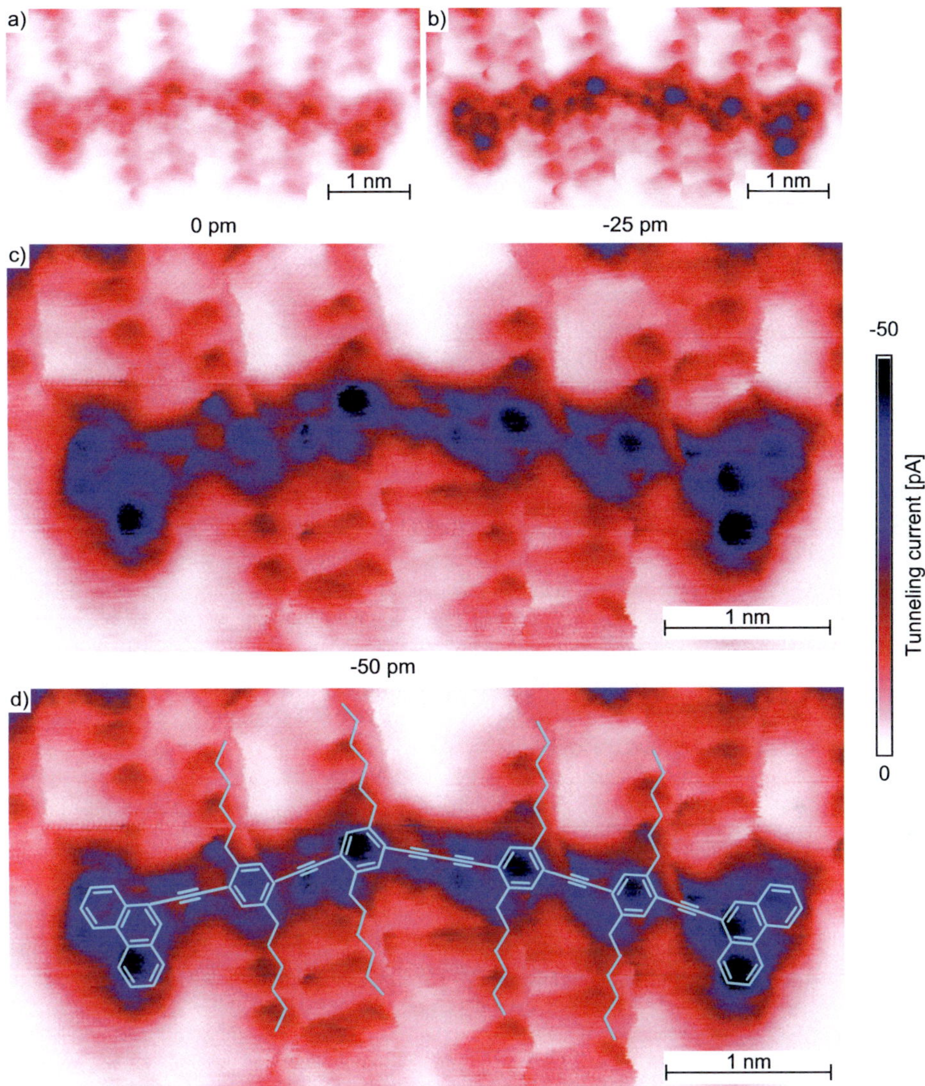

Figure 3.27: Tunneling current images of OPE-B at low bias voltages in constant-height mode, at different tip-sample distances. The chemical structure of the molecule was overlaid in d). Again, individual hexagons of the benzene rings are observed. STM parameters: *Constant-height mode*, $V = -10\,\text{mV}$.

3.6 Conformational Switching of Alkyl Chains

Sequential images of one molecule revealed that the alkyl chains attached to each side of the phenylene groups of OPE-A and OPE-B are not fixed in their position. Instead, switching is observed between two stable conformations. In the topographic image, an alkyl chain either appears higher than the backbone or lower than the backbone, as a line on the surface. For consecutive images of a single OPE-A molecule where this behavior was observed, see Fig. 3.28.

In its pristine state (Fig. 3.28a), the alkyl chains are observed to lie flat on the surface. By performing a detailed scan of the molecule with a higher resolution, thereby increasing the interaction time between STM tip and the molecule, conformational switching can be induced. In Fig. 3.28b four of the alkyl chains already appear higher than be backbone of the molecule. The alkyl chain marked with a red circle switched to its other conformation before Fig. 3.28c was recorded. For the chain marked in green, the switching occurred while scanning the image. Several more transitions of alkyl chains between two conformational states are observed in the following images.

The earliest STM observation of bistable switching was reported by Eigler et al. in 1991, where a Xenon atom moved back and forth between the STM tip and a nickel surface [146]. The transition rate between the two states shows a power-law dependence on the tunneling current, which can be explained by a stepwise vibrational heating due to inelastic electron scattering, where the energy gained from inelastic collisions competes with the energy lost to electron-hole pair generation and phonons [147–149]. Conformational switching of a single organic molecule between two stable states was reported in 2001, where OPEs embedded in a dodecanethiol monolayer were observed to cycle between a low- and a high-conductance state [150]. The switching occurred during scanning, and in some cases could be induced by pulsing the electric field over a molecule.

For the conformational switching of a single, isolated organic molecule a linear dependence of the transition rate on the current was reported, as well as an exponential distribution of residence times in the individual states [53]. Such a behavior is expected for statistically independent one-electron processes [151]. Neglecting temperature and lifetime broadening of the inelastic tunneling rate, the transition rate Γ is then approximately proportional to the conductance G and the effective electron-vibron coupling strength λ [148, 152]:

$$\Gamma_{0/1}(V) \cong \frac{\lambda G}{e}(|V| - V_{\mathrm{th},0/1})\Theta(|V| - V_{\mathrm{th},0/1}), \tag{3.3}$$

with the threshold voltage $V_{\mathrm{th},0/1}$ for switching from state 0 to state 1 and vice versa. With the STM tip held over an alkyl chain in constant-height mode, the tunneling current is observed to alternate between two levels, see Fig. 3.29. Only two stable current levels are observed. The conductance ratio between these two states and the

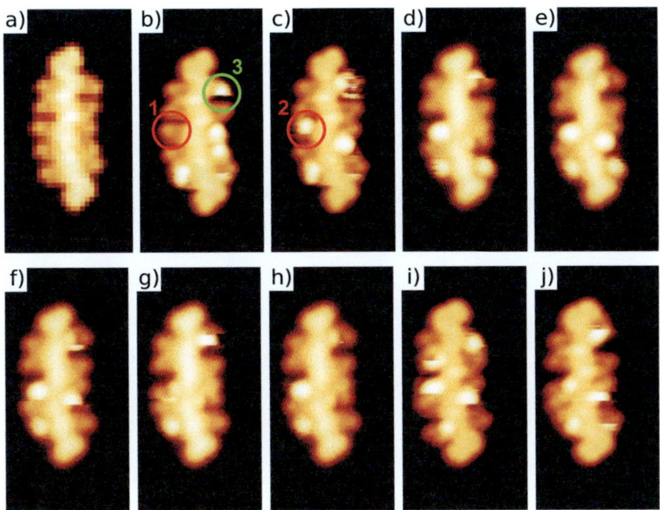

Figure 3.28: Consecutive images of a single OPE-A molecule, where conformational switching of the alkyl chains attached to the phenylene groups is observed. An individual chain appears either lower than the backbone (red circle, 1) or higher than the backbone (red circle, 2). If the switching occurs during scanning, noise is introduced into the image (green circle, 3). STM parameters: $I = 50\,\text{pA}$, $V = -1.5\,\text{V}$.

transition rate depended on the exact position of the tip relative to the molecule. This made spectroscopic measurements over an alkyl chain difficult, as it caused the tunneling current to fluctuate as soon as a certain threshold in bias voltage was reached, see Fig. 3.30. The threshold appeared to be similar for positive and negative bias voltages. To further characterize the switching behavior, height traces over an alkyl chain were recorded in constant-current mode at different tunneling current set points and bias voltages.

Five-second excerpts of two of these traces are plotted in Fig. 3.31a/b, for a bias voltage of -1.5 V and tunneling current set points of 100 pA and 30 pA. Two height levels can be identified in these traces, with a difference of $\approx 70\,\text{pm}$. Consequently, the data was discriminated into two states, 0 and 1. In state 0, the alkyl chain is lying flat on the surface, while in state 1 it appears higher than the backbone. Two qualitative observations can already be derived from this data: The transition rate is higher for a tunneling current of 100 pA than for 30 pA, and the total residence time of the system in state 1 is higher. Histograms of apparent heights for the full traces (each comprising 25000 data points with a time resolution of 20 ms) are shown in Fig. 3.32, where the different occupations of the two states are obvious.

The residence times $\tau_i^{(0/1)}$ in each state were extracted from the time traces and averaged to obtain the mean lifetime of each state $\bar{\tau}^{(0/1)}$:

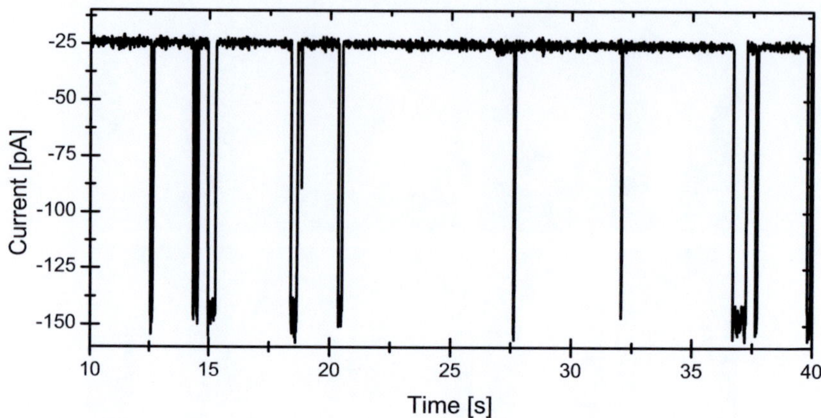

Figure 3.29: With the STM tip positioned over an alkyl chain and constant tip-sample distance, the tunneling current was observed to switch between two levels.

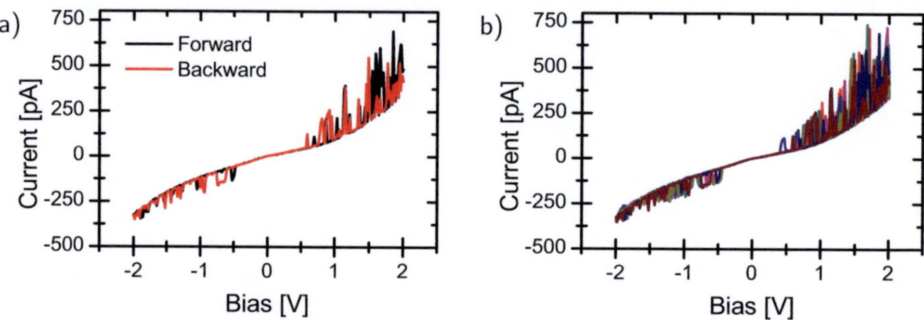

Figure 3.30: a,b) Spectroscopy over an alkyl chain. Switching behavior is observed for both polarities for voltages larger than a certain threshold. The variation in current is greater for positive sample bias.

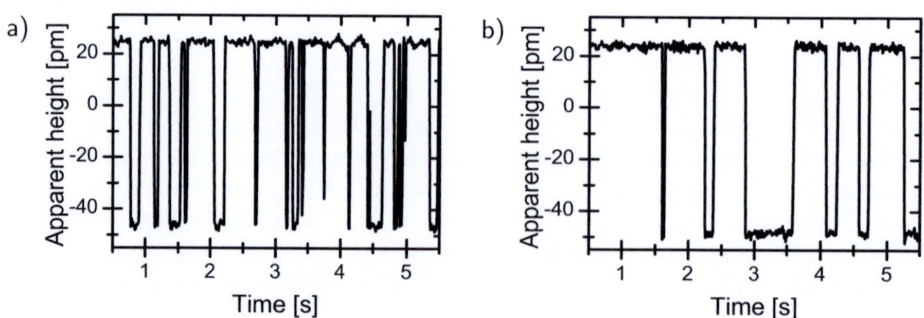

Figure 3.31: a,b) Traces of height vs. time with active feedback loop at a bias voltage of -1.5 V and a set point of a) 100 pA b) 30 pA.

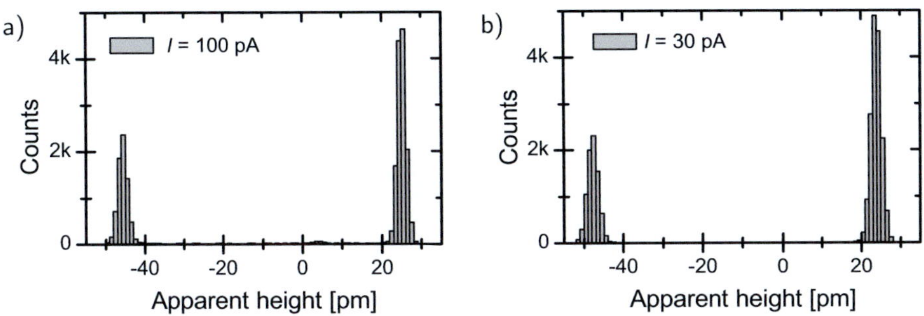

Figure 3.32: a,b) Histogram of apparent heights with active feedback loop at a bias voltage of -1.5 V and a set point of a) 100 pA b) 30 pA.

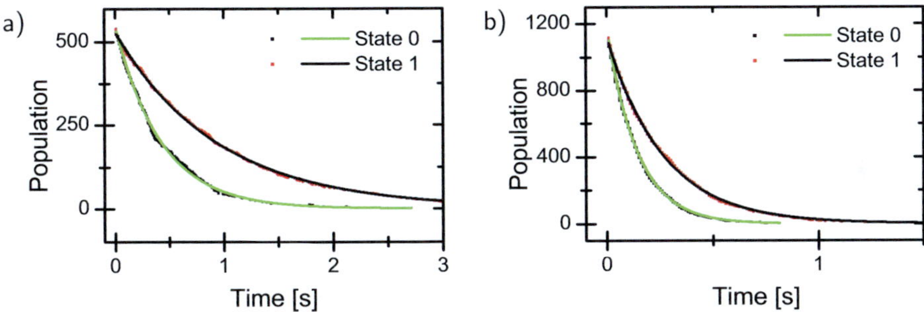

Figure 3.33: a,b) Decay time distribution and exponential fits for a bias voltage of -1.5 V and current set points of a) 30 pA b) 100 pA.

$$\bar{\tau}^{(0/1)} = \frac{1}{N}\sum_{i=1}^{N}\tau_i^{(0/1)}. \tag{3.4}$$

The standard deviation of the mean value $\sigma_{\bar{\tau}^{(0/1)}}$ is taken as the error on each state's lifetime:

$$\sigma_{\bar{\tau}^{(0/1)}} = \sqrt{\frac{\sum_{i=1}^{N}(\tau_i^{(0/1)} - \bar{\tau}^{(0/1)})^2}{N(N-1)}} \tag{3.5}$$

Using the extracted lifetime and the number of switching events as a state's initial population $N_0^{(0/1)}$, exponential decay functions $N^{(0/1)}(t) = N_0^{(0/1)}e^{-t/\bar{\tau}^{(0,1)}}$ can be constructed that fit the data well, see Fig. 3.33.

This analysis was performed for all gathered time traces. A plot of mean lifetimes $\bar{\tau}^{(0,1)}$ vs. current set point is shown in Fig. 3.34. The transition rates $\Gamma_{0/1} = 1/\bar{\tau}^{(0,1)}$

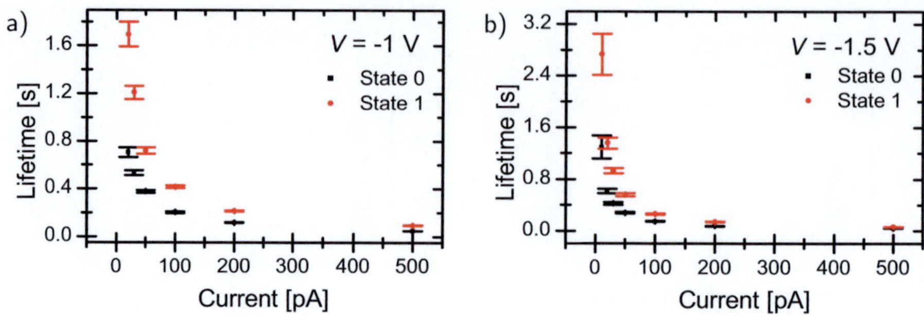

Figure 3.34: a) Lifetime vs. tunneling current for bias voltages of -1 V and -1.5 V.

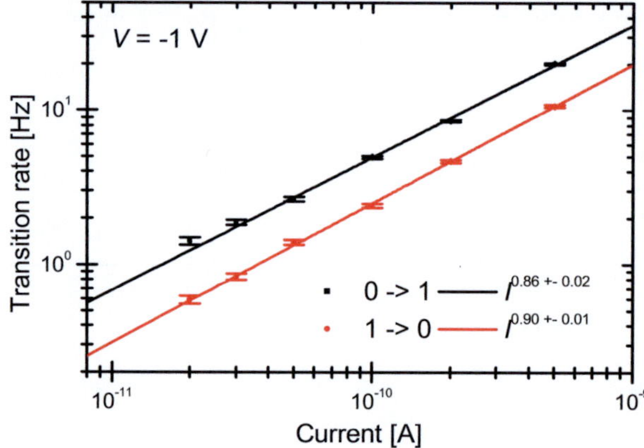

Figure 3.35: Transition rates $\Gamma_{0/1}$ vs. tunneling current for a bias voltage of -1 V.

are plotted vs. current in Fig. 3.35 and Fig. 3.36 in log-log plots. From power law fits with $\Gamma_{0,1} \propto I^n$ exponents of 0.86 ± 0.02 and 0.90 ± 0.01 for a bias of 1 V and 0.82 ± 0.02 and 0.97 ± 0.01 for a bias of 1.5 V are obtained, i.e., the transition rate is roughly proportional to the current. This implies that the transition is a single-electron process, similar to what was reported for conformational switching of single molecules earlier [53]. The energy landscape of the system can be approximated with a double-well potential, as depicted in Fig. 3.37.

From time traces acquired at a constant current of 100 pA and varying bias voltage, we see that the switching rate approaches zero below a certain bias, see Fig. 3.38. The threshold voltages were extracted from linear fits of the low-bias data, including data points up to 1 V. The threshold voltages for the two states are $V_{\text{th},0} = (297 \pm 75)\,\text{mV}$ and $V_{\text{th},1} = (289 \pm 99)\,\text{mV}$. We identify these voltages as the potential-well heights of the double-well potential. These values are not unreasonable for vibronic states of a molecule [153, 154]. Pristine OPE-A molecule are always found in state 0 on

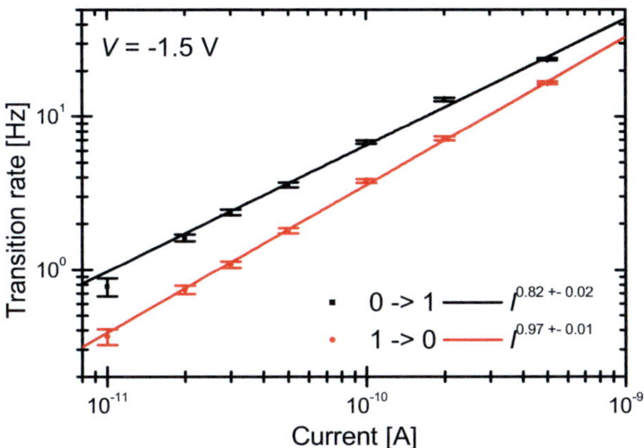

Figure 3.36: Transition rates $\Gamma_{0/1}$ vs. tunneling current for a bias voltage of -1.5 V.

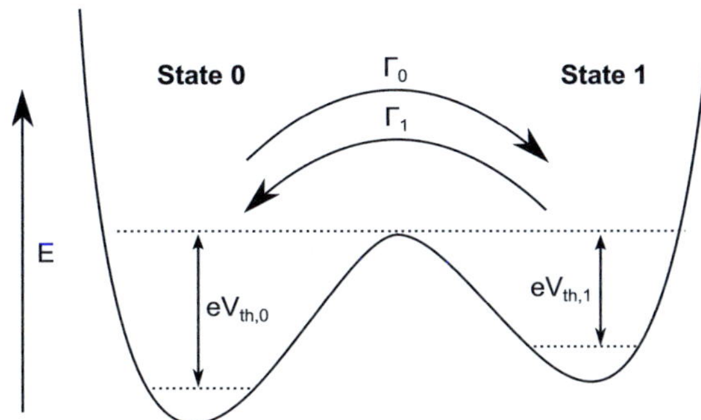

Figure 3.37: Proposed double-well potential of the two-level system of the conformational switching of alkyl chains.

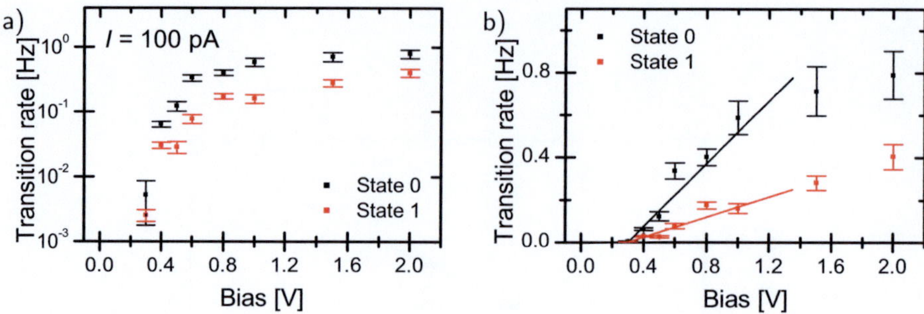

Figure 3.38: Transition rates $\Gamma_{0,1}$ at a fixed current set point of 100 pA. The rate decreases with the bias voltage for both states. Extrapolated switching thresholds using data up to a bias voltage of 1 V are $V_{\mathrm{th},0} = (297 \pm 75)\,\mathrm{mV}$ and $V_{\mathrm{th},1} = (289 \pm 99)\,\mathrm{mV}$.

a freshly-prepared surface, i.e., with their alkyl chains flat on the surface. State 1 is observed only after interaction with the STM tip. Hence, the observed potential energy landscape may only apply for a situation where the STM tip is in close vicinity above the molecule, and otherwise van-der-Waals interactions of the alkyl chains with the Au surface tend to favor state 0 as the preferred conformation. Whilst the observed threshold energies are similar, the error margins are large enough to allow state 0 to lie many kT below state 1, especially at 5 K. The observation bias towards state 0 of pristine molecules may be explained by this energy difference as well.

With bias voltages below 300 mV, switching events could not be observed for reasonable observation times. Indeed, this allows the use of the conformational state as a memory bit: Writing involves waiting for the conformation to switch into the desired state at a higher bias voltage. The state can then be read out at lower bias voltages. Assuming one bit is localized on a $1 \times 1\,\mathrm{nm}^2$ square, this offers a comparably high storage density of $\approx 10^{18}\mathrm{bit/m}^2$ compared to $\approx 10^{16}\,\mathrm{bit/m}^2$ of conventional magnetic hard drives of 2014. However, the use of such a memory storage is of course rather impractical.

3.7 DFT Calculations of OPE-A and OPE-B

In this section, the results of density functional theory (DFT) calculations of the electronic structure of OPE-A and OPE-B, are presented. The results are compared to the STM measurements of both molecules.

All DFT calculations were performed with the program *TURBOMOLE v6.4* [155] on a computer cluster at the Institute of Nanotechnology. To obtain coordinate files for the molecules, their chemical structure was imported into the program *Avogadro* [156]. The geometry was pre-optimized using the built-in Merck Molecular Force Field [157] and then exported to .xyz-format to import it into *TURBOMOLE*. The multipole-accelerated resolution of identity (MARI) approximation was used in all calculations. If not otherwise specified, the self-consistent-field energy was converged to 10^{-6} for structure optimizations, with a grid-size setting of *m4*.

3.7.1 Calculations of Molecules in the Gas Phase

Calculations of both molecules in the gas phase were performed for three different combinations of basis sets and DFT functionals (*levels of theory*). The 6-31g* basis set was chosen because previously published results rely on 6-31g* [36, 39] or on 6-311g [37], which is a triple-valence version of the same. However, in the *TURBOMOLE* community, 6-31g* is tagged as deprecated and not recommended for use. Consequently, calculations were also performed with the def-SVP basis set. It is similar to 6-31g*, except that it also includes polarization functions for hydrogen atoms. For both of these basis sets, the B3LYP functional was used, which is a typical choice for conjugated organic molecules. For later calculations that include one layer of the Au(111) surface, the def-SVP basis set is used exclusively. Furthermore, the BP86 functional was chosen for these calculations because it is computationally less demanding than the B3LYP functional. For comparability the BP86 functional was then included in calculations of molecules in the gas phase as well.

The relaxed geometries of OPE-A and OPE-B in the gas phase are shown in Fig. 3.39. Both molecules relaxed into a planar configuration, with the benzene rings of their phenylene groups in plane. The hexyl chains at the sides of the subunits appear to lie on a comparable flat energy landscape and tend to position themselves freely during geometry optimization.

HOMO, LUMO energies and resulting energy gaps of gas phase calculations of OPE-A and OPE-B are listed in Table 3.1. For both OPE variants, the energy of the HOMO and the LUMO is similar for 6-31g*/B3LYP and def-SVP/B3LYP calculations, with an energy gap of $\approx 2.85\,\text{eV}$ for OPE-A and $\approx 2.92\,\text{eV}$ for OPE-B. This confirms the similarity of the two basis sets. For calculations with the BP86 functional, there are shifts in the energy levels towards each other, with a resulting smaller energy gap of $1.67\,\text{eV}$ for OPE-A and $1.71\,\text{eV}$ for OPE-B. The energy scales of DFT calculations with

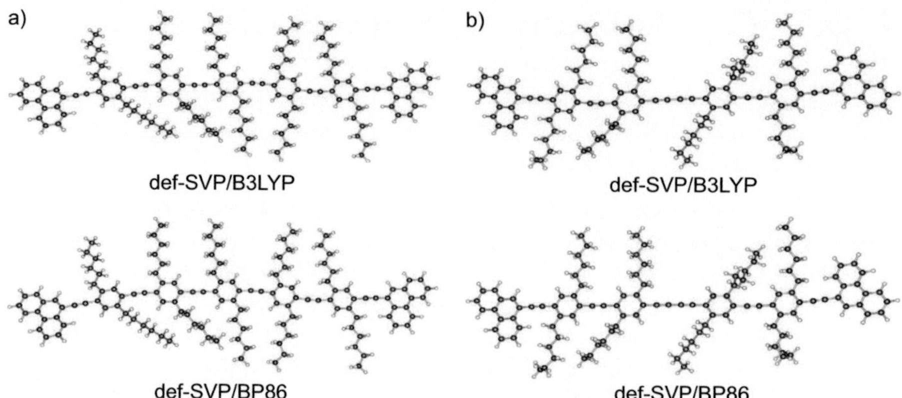

Figure 3.39: Comparison of the relaxed geometry of a) OPE-A and b) OPE-B according to DFT calculations using the B3LYP and the BP86 functionals.

Molecule	Level of theory	HOMO [eV]	LUMO [eV]	E_{gap} [eV]
OPE-A	6-31g*/B3LYP	-4.846	-1.989	2.857
OPE-A	def-SVP/B3LYP	-5.105	-2.259	2.846
OPE-A	def-SVP/BP86	-4.710	-3.037	1.674
OPE-B	6-31g*/B3LYP	-4.966	-2.033	2.933
OPE-B	def-SVP/B3LYP	-5.230	-2.318	2.912
OPE-B	def-SVP/BP86	-4.819	-3.105	1.714

Table 3.1: HOMO, LUMO and energy gaps for both OPE variants in the gas phase according to DFT calculations for different levels of theory.

different functionals are not comparable. However, both calculations yield a slightly larger energy gap for OPE-B compared to OPE-A. For calculations according to def-SVP/B3LYP level of theory, the energy spectrum of OPE-A and OPE-B is plotted in Fig. 3.40 and Fig. 3.41, respectively. For OPE-A (OPE-B), the calculation yields 2610 (2218) molecular orbitals, of which 469 (410) are occupied. Orbital energies corresponding to the highly localized 1s-orbitals of the carbon atoms are not shown in either histogram.

For a graphical representation of the electron density in the HOMO and LUMO of OPE-A and OPE-B, see Fig. 3.42 and Fig. 3.43. These are obtained from the calculations with the def-SVP basis set and the B3LYP and the BP86 functional, respectively. Despite the large difference in energy gap for the calculations with different functionals, the shape of the orbitals is practically identical. The LUMO orbital consists of four lobes per phenylene-ethynylene subunit, two of them located on the top and bottom of the phenylene group. The other two are centered along the axis of the molecule, directly adjacent to the phenylene group. For the HOMO, the calculations yield orbitals with three lobes per phenylene-ethynylene subunit, all centered on the

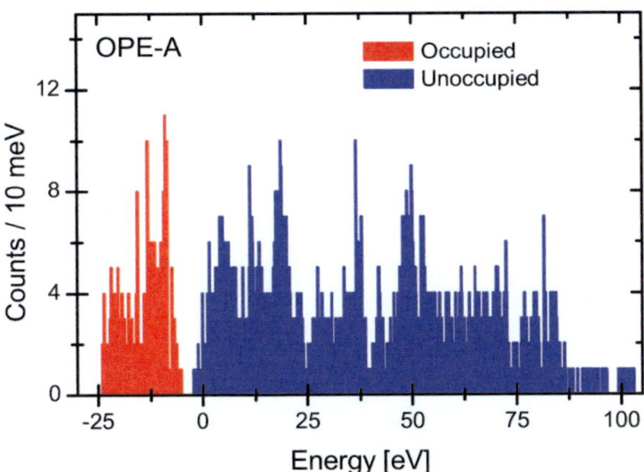

Figure 3.40: Energy spectrum of molecular orbitals of OPE-A according to def-SVP/B3LYP level of theory with a bin size of 10 meV. 130 localized molecular orbitals with energies of $-(277.4 \pm 0.6)$ eV corresponding to the 1s-orbitals of the carbon atoms were omitted.

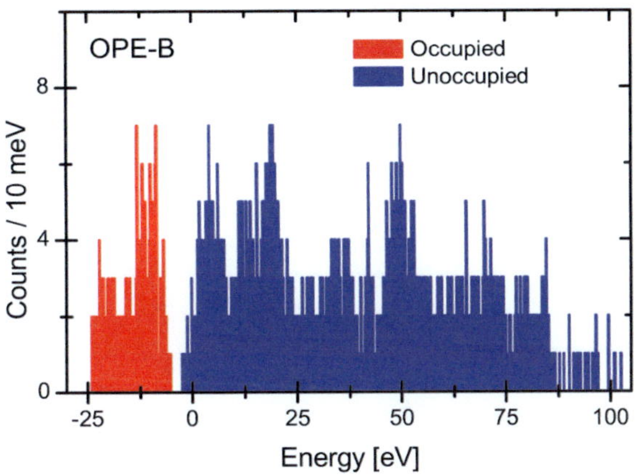

Figure 3.41: Energy spectrum of molecular orbitals of OPE-B according to def-SVP/B3LYP level of theory with a bin size of 10 meV. 112 localized molecular orbitals with energies of $-(277.4 \pm 0.6)$ eV corresponding to the 1s-orbitals of the carbon atoms were omitted.

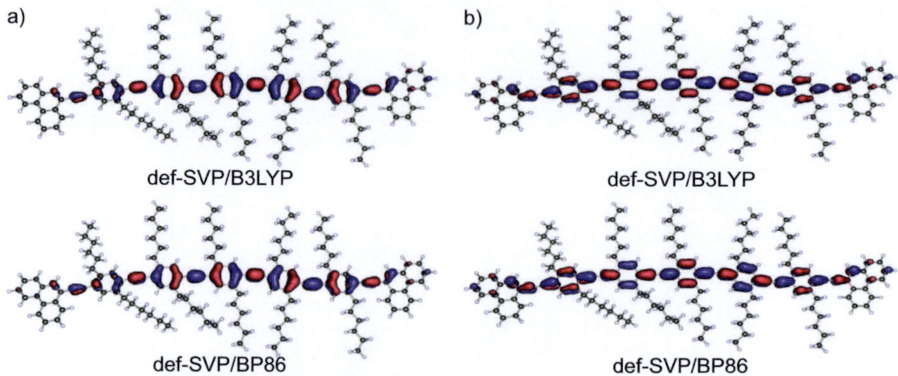

Figure 3.42: Comparison of the a) HOMO and b) LUMO orbital of OPE-A according to DFT calculations using the B3LYP and the BP86 functionals. In this and all following plots of molecular orbitals, red and blue surfaces indicate surfaces of constant wave function amplitude of $A_{\text{red}} = -A_{\text{blue}} = 0.25|A_{\text{max}}|$.

long axis of the molecule. The two lobes on the phenylene group, located on its left and right side, are more elliptical than the third lobe sitting on the center of the ethynylene group.

Most previous reports of DFT calculations of OPEs included at most three phenylene groups [36, 39, 40]. However, the reported shapes of the LUMO and HOMO orbitals are merely shortened versions of the orbital shapes Lu et al. [25] and we have calculated. It appears that the phenylene-ethynylene subunits of an OPE are relatively independent of each other in the DFT calculation, and hence the electronic structure of a planar elongated OPE can be derived from calculations of a shorter one.

As reported in the literature (see Sec. 2.1), substituents of OPE molecules can profoundly influence the orbital structure of the molecule [39]. To find out whether the hexyl chains protruding from the sides of our OPE variants play a significant role, DFT calculations of OPE-A and OPE-B were repeated with the alkyl chains replaced by methyl groups. See Table 3.2 for the calculated HOMO, LUMO levels and the resulting energy gaps. While the absolute values of the energy levels are shifted by up to 2 %, the energy gaps are identical within the error margins of the calculations. The resulting shapes of the LUMO and HOMO are identical to those of the molecules with hexyl chains as well, see Fig. 3.44.

3.7.2 Comparison with STM Measurements

For def-SVP/B3LYP level of theory, the DFT calculations yield energy gaps of 2.85 eV for OPE-A and 2.91 eV for OPE-B. This is in good agreement with the energy gaps observed in optical fluorescence measurements of 2.95 eV for OPE-A and 2.90 eV for

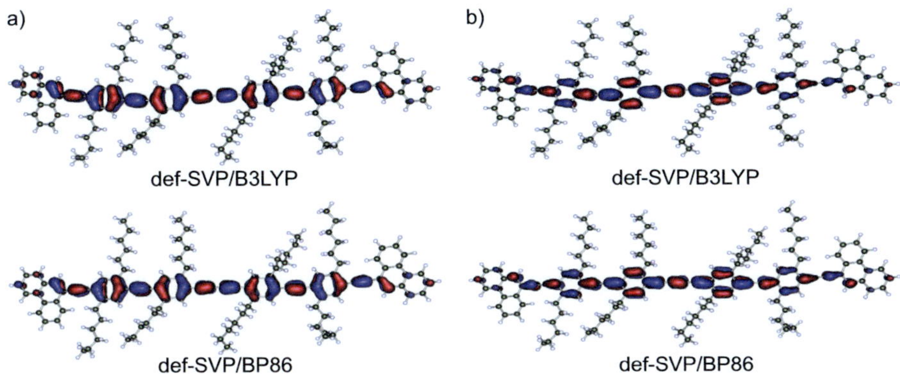

Figure 3.43: Comparison of the a) HOMO and b) LUMO orbital of OPE-B according to DFT calculations using the B3LYP and the BP86 functionals.

Molecule	Level of theory	HOMO [eV]	LUMO [eV]	E_{gap} [eV]
OPE-A*	6-31g*/B3LYP	-4.917	-2.060	2.857
OPE-A*	def-SVP/B3LYP	-5.170	-2.316	2.854
OPE-A*	def-SVP/BP86	-4.773	-3.091	1.682
OPE-B*	6-31g*/B3LYP	-5.015	-2.078	2.936
OPE-B*	def-SVP/B3LYP	-5.265	-2.351	2.914
OPE-B*	def-SVP/BP86	-4.855	-3.137	1.717

Table 3.2: HOMO, LUMO and energy gap for both OPE variants, with hexyl chains replaced by methyl groups.

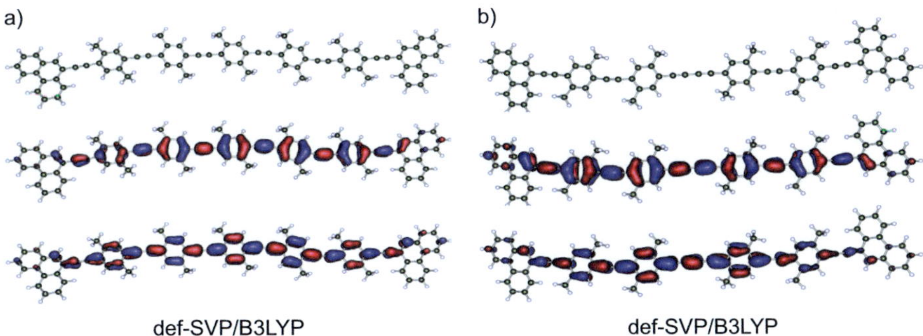

Figure 3.44: Geometry, HOMO and LUMO of a) OPE-A* and b) OPE-B* with hexyl chains replaced by methyl groups.

OPE-B, which are presented in Sec. 2.1.1. In spectroscopic STM measurements presented in Sec. 3.4, a HOMO-LUMO gap of $\approx$4.4 eV was estimated for OPE-A and $\approx$3.2 eV for OPE-B. An energy gap of $\approx$3 eV for an OPE of similar length was also reported by Lu et al[25], measured by ultraviolet-visible spectroscopy. The gap obtained here from DFT calculations is in good agreement with these values and the fluorescence measurements. The values obtained by STM spectroscopy are likely to be overestimated, due to the strong hybridization of the LUMO with states of the conductive surface, which made it difficult to identify peaks in the density of states at positive bias voltages. A direct comparison of near-gap orbital energies according to DFT calculations with the spectroscopic data is shown in Fig. 3.45 for OPE-A and Fig. 3.46 for OPE-B. In both cases the energies according to DFT calculations were shifted, by 3.41 eV (OPE-A) and 3.68 eV (OPE-B), to align the assumed HOMO level of the spectroscopic data to the energy scale of the DFT calculation. The HOMO was chosen for this alignment because measurements have shown that the Au surface states more strongly hybridize with the LUMO, and therefor the position of the HOMO is assumed to be less influenced by the fact that the DFT calculations were of the molecules in gas phase. For OPE-A, the first unoccupied orbital then lies in a region where the density of states (DOS) from spectroscopic data merely increases. At its peak at $\approx$2.8 eV, many molecular orbitals are already found according to the DFT results. For OPE-B, the agreement between the DFT calculation and the spectroscopic data is much better. The energetic position of the four highest occupied molecular orbitals corresponds roughly to peaks in the DOS of the molecule. For unoccupied molecular orbitals, the correlation is not as pronounced, but the energy gap of OPE-B seems to be reproduced reasonably well by the DFT calculation.

For the comparison of the calculated shapes of the HOMO and LUMO orbitals of OPE-A and OPE-B with the conductance maps shown in Sec. 3.4.1, we again choose the results from the def-SVP/B3LYP level of theory. In any case, the calculated orbital shapes do not depend on the level of theory of the DFT calculation. A side-by-side comparison of the DFT calculations and a conductance map of an OPE-A molecule is shown in Fig. 3.47. The recorded conductance map of the HOMO corresponds very well to the DFT calculations. Three lobes per phenylene-ethynylene subunit are observed in both the conductance map and the electron density according to the DFT calculation. However, the different elliptical shapes of the individual lobes are not observed in the conductance map. The STM conductance map obtained from the LUMO is not recognizable in the corresponding electron density plot according to the DFT calculation. At even lower bias voltages, it was possible to record the shape of the HOMO-1 orbital with the STM. In the DFT calculation, this shape can be constructed by the sum of four orbitals with an energy of 1.52 eV to 1.63 eV below the HOMO level. A side-by-side comparison of the DFT results and the conductance map is shown in Fig. 3.48. The map of HOMO-1 was recorded at -2.5 V, with the same shape of the HOMO recorded from -2 V to -1 V. Hence, the energy difference between HOMO and HOMO-1 in the STM conductance map and the DFT calculation is similar.

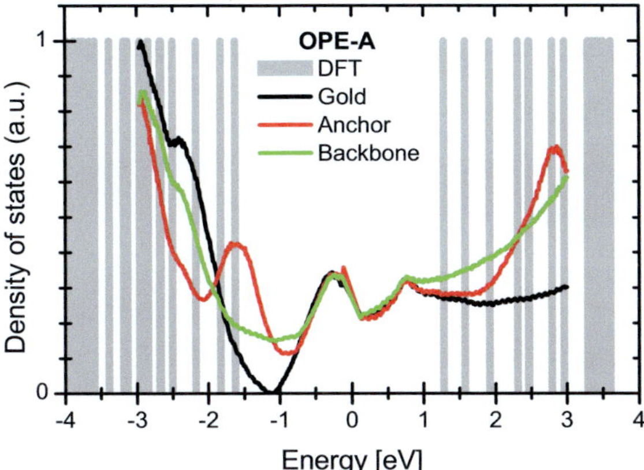

Figure 3.45: Comparison of STS measurements and results of DFT calculation according to def-SVP/B3LYP level of theory for OPE-A. Plotted is the averaged $(\mathrm{d}I/\mathrm{d}V)/(I/V)$ signal over various positions of the molecule. Gray bars correspond to energies of molecular orbitals from the DFT calculation. The DFT energies were shifted by 3.41 eV to align the HOMO level with the assumed HOMO level of the STS spectra.

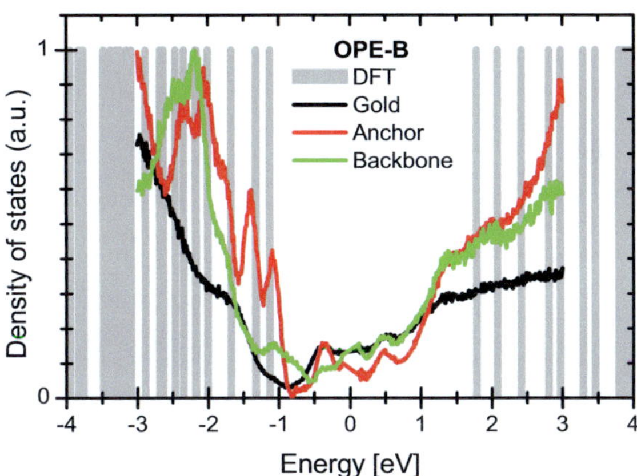

Figure 3.46: Comparison of STS measurements and results of DFT calculation according to def-SVP/B3LYP level of theory for OPE-B. Plotted is the averaged $(\mathrm{d}I/\mathrm{d}V)/(I/V)$ signal over various positions of the molecule. Gray bars correspond to energies of molecular orbitals from the DFT calculation. The DFT energies were shifted by 3.68 eV to align the HOMO level with the assumed HOMO level of the STS spectra.

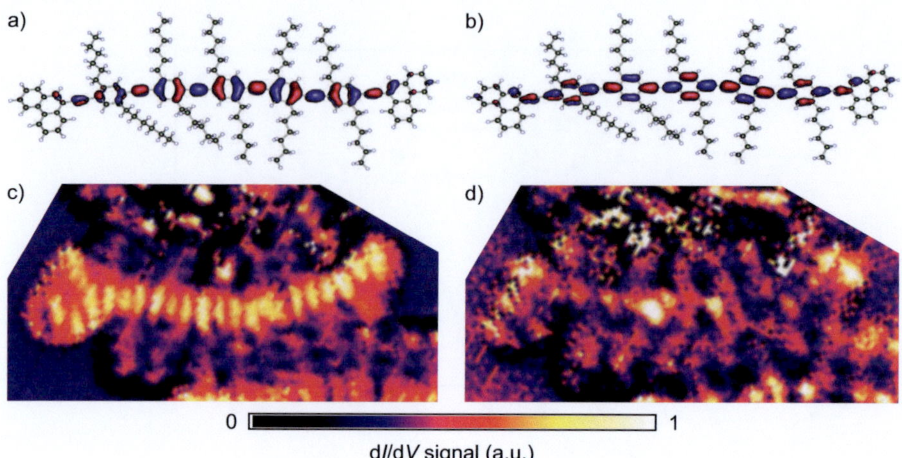

Figure 3.47: a) HOMO (-5.11 eV) b) LUMO (-2.26 eV) of OPE-A according to DFT calculations using def-SVP/B3LYP level of theory. c,d) Corresponding STM conductance maps of an OPE-A molecule at bias voltages of c) -1.5 V d) 2 V. STM parameters: $I = 200\,\text{pA}$, $f_{osc} = 1.33\,\text{kHz}$, $A_{osc} = 42\,\text{mV}$.

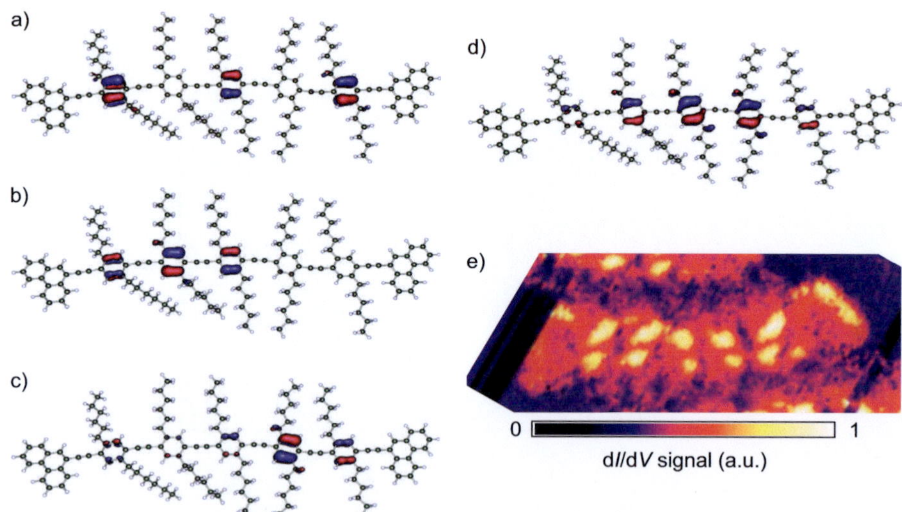

Figure 3.48: a-d) Orbitals 458-461 of OPE-A according to the DFT calculation using def-SVP/B3LYP level of theory, which resemble the observed STM conductance map of the HOMO-1 orbital recorded at a bias voltage of -2.5 V shown in e). DFT energies: a) Orbital 458, -6.735 eV b) Orbital 459, -6.713 eV c) Orbital 460, -6.708 eV d) Orbital 461, -6.626 eV. e) STM parameters: $I = 200\,\text{pA}$, $f_{\text{osc}} = 1.33\,\text{kHz}$, $A_{\text{osc}} = 42\,\text{mV}$.

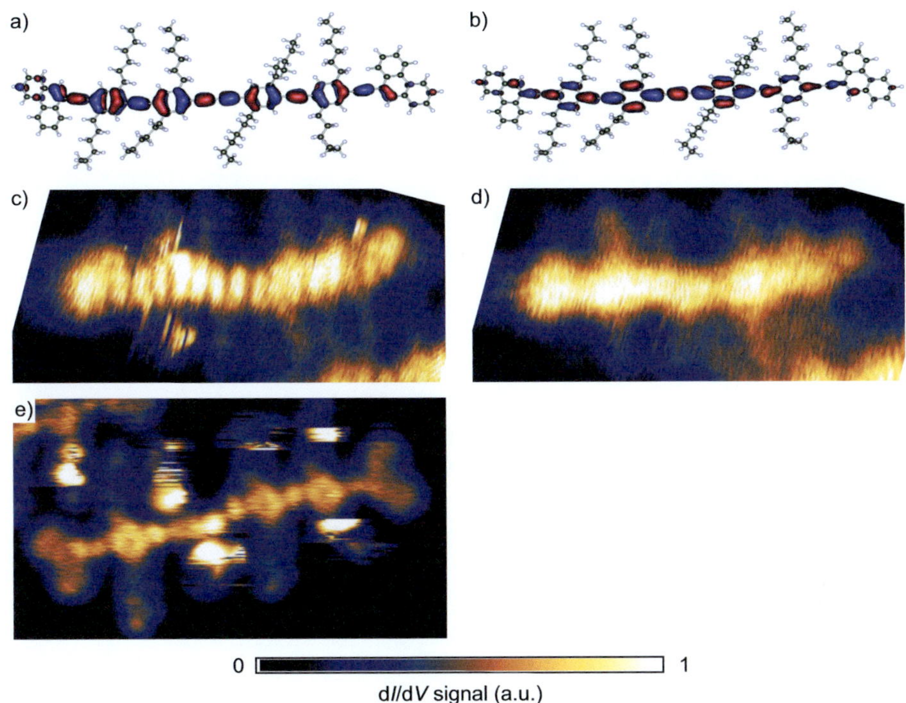

Figure 3.49: a) HOMO (-5.23 eV) b) LUMO (-2.32 eV) of OPE-B according to DFT calculations using def-SVP/B3LYP level of theory. c,d,e) Corresponding STM conductance maps of an OPE-B molecule at bias voltages of c) -1.5 V d) 1.5 V e) -1.5 V. STM parameters: *Constant-height mode*, $f_{\mathrm{osc}} = 8.16\,\mathrm{kHz}$, $A_{\mathrm{osc}} = 100\,\mathrm{mV}$.

For the OPE-B molecule, a side-by-side comparison of the recorded conductance maps of the LUMO and HOMO with DFT calculations from def-SVP/B3LYP level of theory is shown in Fig. 3.49. The resolution of the LUMO conductance map is even poorer in this case, allowing no clear distinction of orbital lobes. At a bias voltage of -1.5 V, two conductance maps are shown, recorded with different STM tips. The orbital shape shown in Fig. 3.49c resembles the shape recorded for OPE-A, shown in Fig. 3.47c. The same number of lobes are observed, three per phenylene-ethynylene subunit. The conductance map shown in Fig. 3.49e additionally shows the different elliptical shapes that are found in DFT calculations of the HOMO, resembling the calculation very well.

A probable cause for the difficulty in resolving the LUMO orbital in the STM conductance maps might be hybridization with the Au(111) surface, which also prevented the identification of clear peaks in the spectroscopic measurements for positive bias voltages. To try and account for hybridization effects in the DFT calculations, one layer of the gold surface was subsequently included in the calculation.

3.7.3 Calculations including the Au(111) Surface

To improve the comparability with STM conductance maps, a closely-packed layer of gold atoms with a nearest neighbor spacing of 2.89 Å, representing an unreconstructed Au(111) surface, was included in the DFT calculation. To reduce the number of required surface atoms, the OPE-A molecule was shortened to two phenylene-ethynylene subunits. As we have seen from calculations of shorter molecules in the literature, the orbital shapes per individual phenylene-ethynylene subunit are identical to that of longer OPEs. The short OPE-A variant shall be referred to as OPE-A2. A slab of 11×14 gold atoms then provided sufficient surface area. To further reduce the computational requirements, the BP86 functional and the def-SVP basis set were chosen. Sixty of the inner-shell electrons of the gold atom are approximated by an electronic core potential taken from Andrae, Häußermann, Dolg, Stoll and Preuß [158]. The outer electrons of each gold atom are represented by the 27 basis functions of the def-SVP basis. For molecules on inert surfaces like gold, where the formation of covalent bonds is unlikely, van-der-Waals interactions can become crucial. They were incorporated into all calculations by the DFT-D3 approximation with Becke-Johnson damping [159, 160].

An OPE-A2 molecule was loosely placed above the slab of gold atoms and the geometry relaxation started. Eventually, the molecule relaxed into a planar configuration on the surface, as shown in Fig. 3.50. In our STM study, we almost always have observed this planar configuration for both OPE-A and OPE-B. The backbone of the molecule is found to be slightly bent, which we have also observed often in our STM study. It appears to be induced by an attractive interaction of neighboring alkyl chains.

The shape of the HOMO orbital for OPE-A2 on Au(111), shown in Fig. 3.52, represents a shortened version of the HOMO of OPE-A, as expected. As the calculation does not treat orbitals of the surface and of the molecule differently, the calculation yields almost no energy gap for the entire system, as expected for a metallic surface. To find out whether molecular orbitals have hybridized with the surface or not, each individual orbital of the complete system has to be inspected and its localization analyzed.

The lowest unoccupied orbital of the molecule-surface system is depicted in Fig. 3.51a. The electron density is clearly localized around different atoms of the surface. The lowest unoccupied orbitals that are localized on the OPE-A2 molecule are shown in Fig. 3.51b-d (Orbitals 1745, 1747, 1748). Their shapes resemble that of a LUMO orbital of the OPE molecules in the gas phase. The orbitals display varying degrees of hybridization with surface states, Fig. 3.51a/b more so than Fig. 3.51c. Their energies range from -3.68 eV to -3.48 eV. On the other hand, the highest occupied orbital of the molecule-surface system is shown in Fig. 3.52. The electron density is localized on the OPE-A2 molecule and the shape is similar to the HOMO of OPE-A obtained from DFT calculations and from STM observations. It lies at an energy of -5.52 eV, leading to an energy gap of the molecule on the surface of ≈ 1.94 eV.

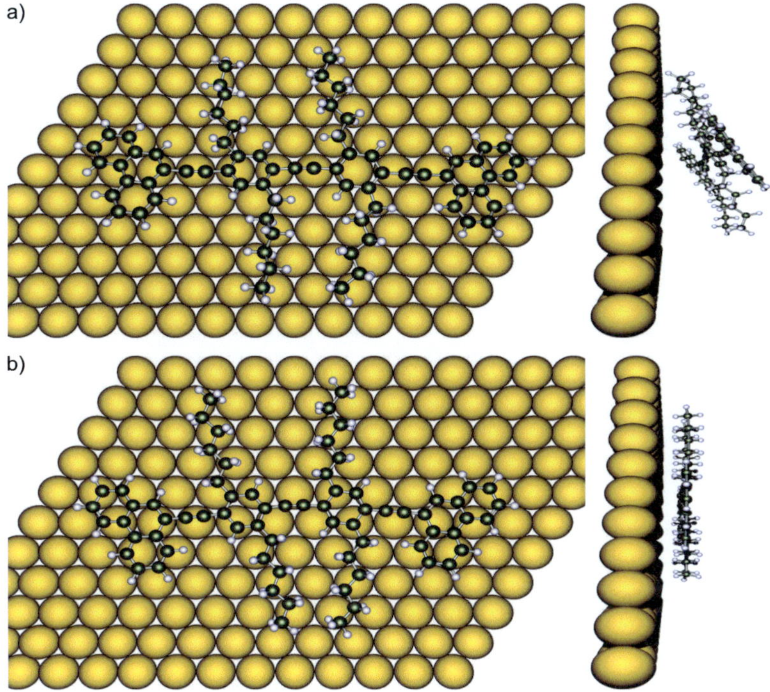

Figure 3.50: Top and side views of the geometry of OPE-A2 on one layer of Au(111) a) before and b) after relaxation.

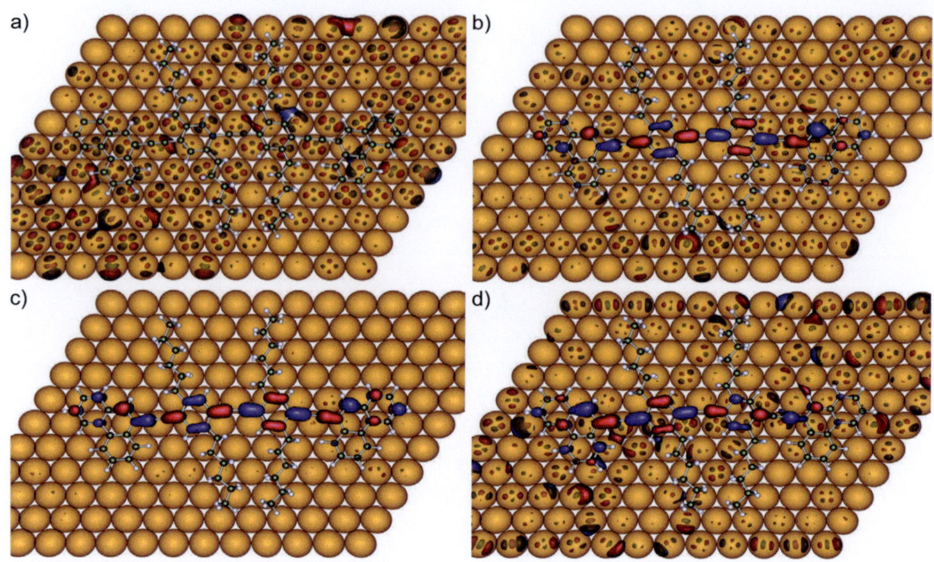

Figure 3.51: a-d) Molecular orbitals resembling the typical shape of unoccupied orbitals of OPE molecules in gas phase calculations, on one layer of Au (111). a) Orbital 1711, -5.461 eV b) Orbital 1745, -3.728 eV c) Orbital 1747, -3.676 eV d) Orbital 178, -3.475 eV.

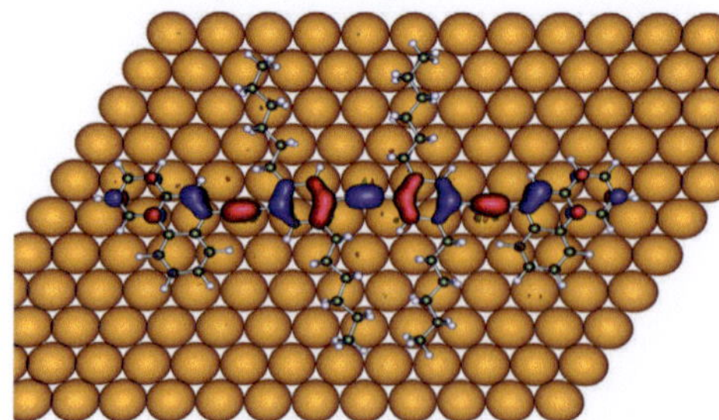

Figure 3.52: HOMO of OPE-A2 on one layer of Au(111). Orbital 1710, -5.524 eV.

Molecule	**HOMO [eV]**	**LUMO [eV]**	E_{gap} **[eV]**
OPE-A2	-4.963	-1.888	3.07
OPE-A2 on Au(111), 1 layer	-5.524	≈-3.58	≈1.94

Table 3.3: HOMO, LUMO and energy gap of OPE-A2 in the gas phase and on a single layer of an unreconstructed Au(111) surface.

For comparison of the energy scale of the molecule on the surface with a molecule in the gas phase, a single-point calculation of the relaxed surface geometry was run without the surface slab, see Fig. 3.53 for the results. The shapes of the LUMO and HOMO are reminiscent of those of OPE-A, as expected. The energy gap of OPE-A2 in gas phase is 3.07 eV, which is more than 1 eV higher than that of the molecule on the surface. The HOMO orbital shifted to -4.96 eV, which is just ≈ 560 mV higher than the HOMO energy of the molecule on the surface. However, the LUMO shifted by ≈ 1.7 eV, underlining the strong hybridization with the gold surface. All calculated energies are listed in Table 3.3. The pronounced energy shift of the LUMO orbital on the surface might as well explain why it is difficult to image the LUMO with the STM as well: Strong hybridization with surface states typically inhibits the direct imaging of molecular orbitals, as described in Sec. 2.2.1. Calculations with the BP86 functional yielded an energy gap of 1.67 eV and 1.71 eV for OPE-A and OPE-B, which is in both cases too small by about 1.3 eV. For OPE-A2 in the gas phase, an energy gap of 3.07 eV is found. Lu et al. also found that the energy gap of OPEs decreases with their width [25]. Due to the use of the BP86 functional in our calculations of OPE-A2, the energy gap of OPE-A2 is likely to be underestimated. Still, in our STM measurements of OPE-A and OPE-B on a Au(111) surface a decreased energy gap compared to the molecule in solution is not observed. The trustworthiness of these DFT results with the Au surface might be increased if more than one layer were included in the calculation, or the B3LYP functional were used. Unfortunately, the computational requirements needed for these options are out of our reach.

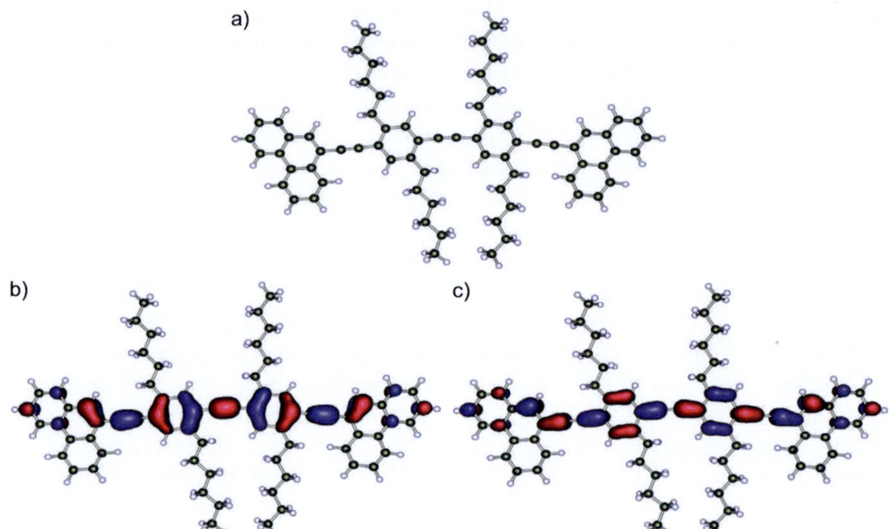

Figure 3.53: a) Relaxed geometry of OPE-A2, taken from the calculation that included a slab of the Au(111) surface. b) HOMO c) LUMO of OPE-A2 in gas phase from a single-point calculation.

4 Carbon Nanotube Nanogap Electrodes

After completing the STM characterization of the OPE molecules, direct electrical measurements were attempted by placing the molecules into a carbon nanotube nanogap embedded in a device geometry. To that end, two different fabrication routes of CNT nanogap electrodes were explored, which are presented in this chapter. The results pertaining to nanogap fabrication by electron-beam-induced oxidation (EBIO) have been published in [3]. Furthermore, in the course of this PhD work, EBIO was successfully used to pattern flakes of graphene, too, which has been published in [2]. In the second part of this chapter, nanogap fabrication by helium-ion sputtering is presented, which has been published together with first results on embedding OPE-A molecules into nanogap electrode devices in [1].

4.1 Nanogaps by Electron-Beam-Induced Oxidation

4.1.1 Sample Preparation

Three kinds of devices were fabricated to explore electron-beam-induced oxidation (EBIO) of carbon nanotubes on different scales and different substrates: Single-CNT and thin-film devices on Si/SiO_x substrates and single-CNT devices on free-standing silicon nitride membranes. For a side view of both types of devices, see Fig. 4.1. Suspensions of metallic single-walled carbon nanotubes with a diameter of (1.2 ± 0.2) nm were prepared by Dr. Frank Hennrich by the method described in Sec. 2.4.2.

Devices on Silicon Oxide

Thermally oxidized p-doped silicon wafers were used both for single-CNT and for thin-film devices. In both cases the oxide thickness was 800 nm. The silicon substrate was heavily p-doped, which allowed its use as a backgate. Sputtered metal electrodes made of 5 nm titanium and 50 nm palladium were fabricated onto the substrates by standard electron-beam-lithography as described in Sec. 2.4.4. The layout is depicted in Fig.

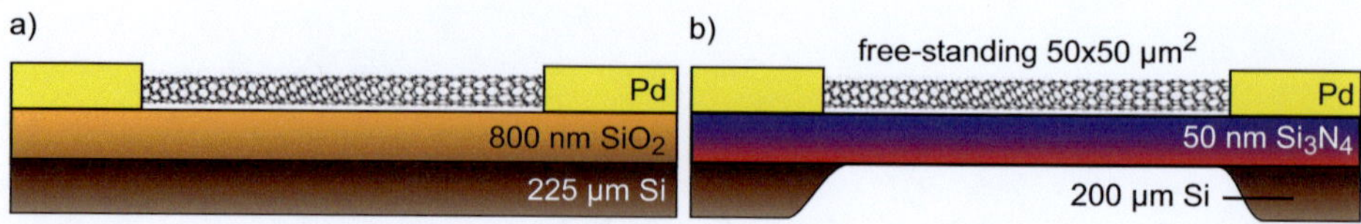

Figure 4.1: Cross-sectional view of devices on a) SiO_2 and b) free-standing Si_3N_4 membranes. Metallic single-walled carbon nanotubes are contacted by lithographically defined palladium contacts spaced 700 nm apart.

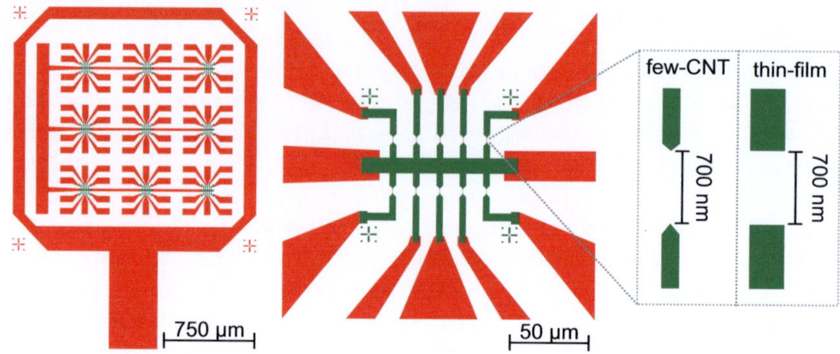

Figure 4.2: Lithographic mask for metal electrodes on oxidized silicon substrates. Electrode pairs are grouped into nine sets of ten each. For few-CNT devices, the electrodes have sharp tips, while for thin-film devices they are flat. In both cases the gap between drain and source electrodes is 700 nm wide. Structures colored in green are written with a 20 µm aperture and structures colored in red with a 120 µm aperture.

4.2. The electrode pairs were placed 700 nm apart and were ≈ 200nm wide and had a sharp tip for few-CNT devices, whereas for thin-film devices, they were ≈ 500 nm wide and flat. The electric field strength is highest for sharp electrodes and more uniform for flat electrodes, favoring single CNTs for the former kind and parallel alignment of multiple CNTs for the latter kind. Ten electrode pairs were grouped into each 100x100 μm^2 field, which was connected to a larger super-structure with contact pads for the common drain electrode and each individual source electrode. The 100x100 μm^2 fields were written with an aperture size of 20 µm, whereas the larger structures were written with a 120 µm aperture. The smaller aperture leads to longer writing time but also to a better resolution, allowing the fabrication of the sharp electrode tips.

Metallic single-walled CNTs were deposited onto the metal contacts by dielectrophoresis, as described in Sec. 2.4.3. An alternating voltage with a frequency of 300 kHz was applied between the backgate and the common drain electrode. The source electrodes were capacitively coupled to the backgate and it was hence not necessary to connect them individually. A drop of nanotube dispersion with a concentration of ≈ 5 nanotubes per μm^3 was placed onto the device for three minutes. After this time, it was first diluted with water, then with methanol and finally blow-dried with nitrogen. For

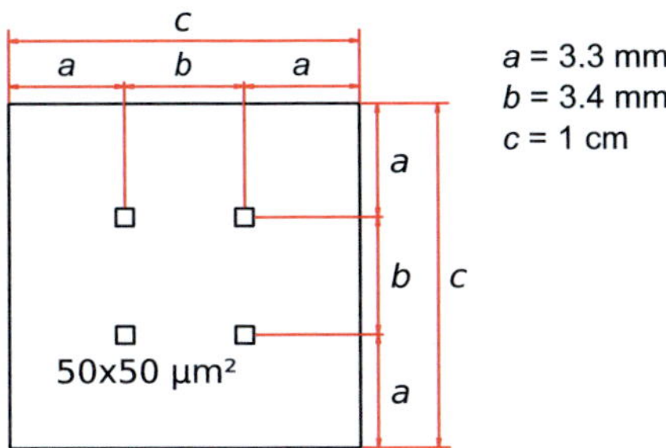

Figure 4.3: Schematic top view of the 1x1 cm^2 silicon nitride membrane substrate. The substrate is uniformly covered with 50 nm of silicon nitride and has been etched from the underside in four positions to yield free-standing silicon nitride membranes.

few-CNT devices, a peak-to-peak voltage of $V_{pp} = 1.5$ V was used, whereas for thin-film devices, a discontinuous alternating voltage with $V_{pp} = 6$ V and $\tau_{on} = 10$ ms and $\tau_{off} = 90$ ms was used to prevent electroosmosis as explained in [161].

Devices on Silicon Nitride Membranes

For the second kind of samples, commercially available silicon nitride membranes as depicted in Fig. 4.3 were used. A 50 nm thin silicon nitride film is supported on 1x1 cm^2 silicon substrates. On each 1x1 cm^2 substrate the silicon had been completely etched from the bottom side in four spots of 50x50 μm^2, producing four free-standing silicon nitride membranes.

On each of these four membranes, metal contacts were fabricated of 5 nm titanium and 50 nm palladium. The electron-beam lithography layout is depicted in Fig. 4.4. The lithography had to be aligned to within ± 5 µm, so that the central part of the design was lying on top of the free-standing silicon nitride. The electron dose for photoresist exposure had to be increased on the membrane because of the lack of backscattered electrons.

For the dielectrophoresis, an alternating voltage with $V_{pp} = 1.5$ V and a frequency of 300 kHz was used. Each source electrode had to be contacted individually for the deposition, as there was no backgate that could be used for capacitive coupling. This was achieved either using multiple probe needles of a probe station, or by wire-bonding the source electrodes to a single contact pad before deposition. Otherwise, the same nanotube dispersion and deposition procedure as for few-CNT devices on the silicon oxide was used.

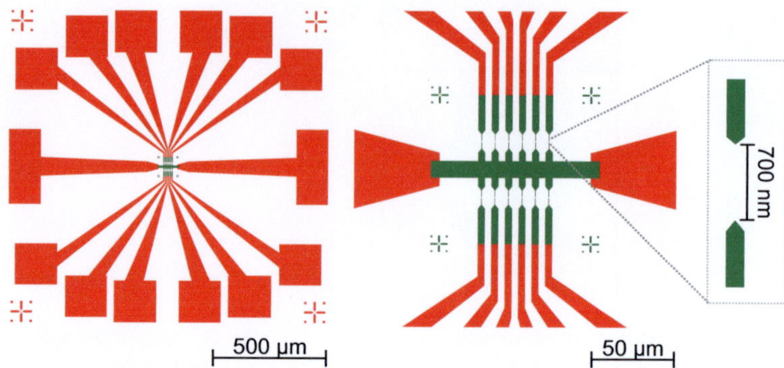

Figure 4.4: Lithographic mask for metal electrodes on silicon nitride membranes. The center part consists of ten electrode pairs, and is aligned during lithography to reside on the free-standing silicon nitride membrane.

4.1.2 Experimental Setup

Imaging and cutting of nanotube devices was carried out in a Zeiss Ultra Plus electron microscope. The built-in charge compensation needle allows the injection of oxygen gas (purity 99.998 %) into the chamber close to the substrate. The tip of the needle was adjusted at 200 µm off the center of the electron-beam scan window, while its vertical position was fixed to less than 50 µm above the device surface, by touching down on it and then lifting it just above the substrate.

The gas flow was regulated by a mass-flow controller and set to (35 ± 2) sccm/min. The main chamber of the microscope was continuously pumped during the experiment with a turbomolecular pump. During gas injection the system pressure read as $5.7 \cdot 10^{-3}$ mbar, with the local pressure close to the surface at the center of the scan window presumably higher. The molecular flux at the exit of the needle was calculated to $8 \cdot 10^{7}\,\mathrm{nm^{-2}s^{-1}}$.

The SEM is also equipped with two Kleindiek MM3A-EM nanoprobers, which can move freely inside the microscope chamber. The microscope stage, which is electrically isolated from ground and electrically accessible by a vacuum feed-through into the chamber, was used as a third electrical contact. This allowed simultaneous control of source-drain and gate voltage. All electrical measurements were performed with a dual-channel Keithley 2636A source/measure unit.

4.1.3 Electrical Characterization and Conditioning

To quickly assess which of the electrode pairs were bridged by carbon nanotubes, voltage-contrast SEM was used [162]. The contrast of the source electrode is observed while a voltage is applied between the common drain electrode and the backgate. If

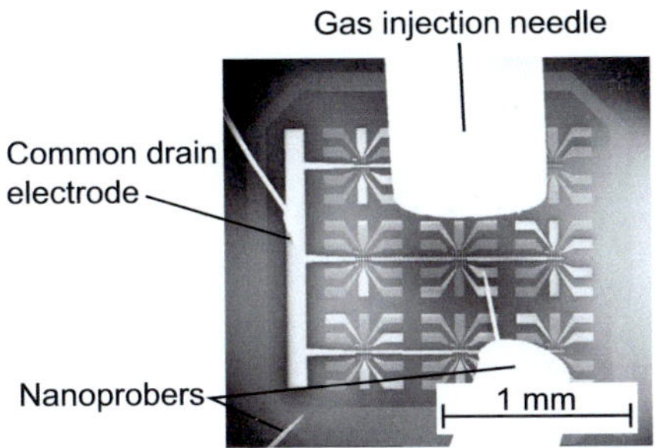

Figure 4.5: The experimental setup on silicon oxide as seen in the electron microscope. The common drain electrode is wire-bonded to the stage, which is electrically accessible from the outside through a vacuum feed-through. The two nanoprobers are used to contact an individual carbon nanotube's source electrode and the gate electrode on the substrate, respectively.

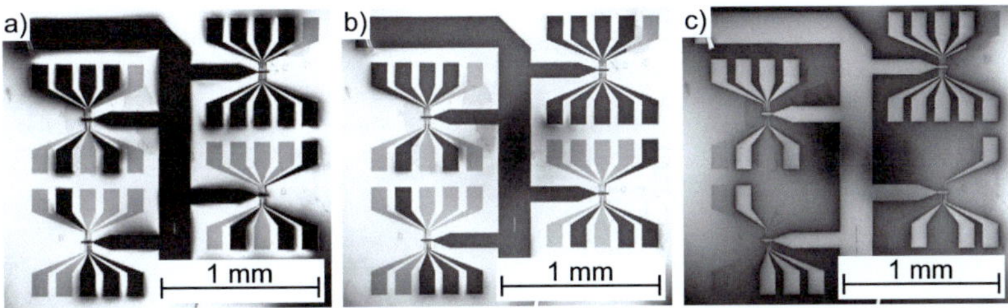

Figure 4.6: Voltage-contrast SEM for metallic contacts for a) 15 V b) 0 V and c) -15 V applied between drain electrode and backgate. For the devices with a metallic carbon nanotube between electrode pairs, the source electrode contact pad follows the contrast of the common drain electrode. Contrast of other (electrically) floating source contact pads does not change significantly with the applied voltage.

a metallic CNT bridges the source and drain electrodes, their contrast is expected to be identical, because there is no significant difference in their electric potential. SEM images of metallic electrodes where some are bridged by metallic CNTs and observed under different applied voltages are shown in Fig. 4.6. When a positive voltage is applied to the drain electrode, less secondary electrons are able to escape from the surface, as they are attracted by the electric field. Hence the electrodes appear darker. For negative voltages, the opposite is the case. Electrically floating source electrodes, which are not connected to the drain by a conductive channel, change their contrast mainly due to charge effects.

As-prepared CNT devices exhibited a conditioning effect: Their resistance decreased permanently when a high current was applied for the first time. The dissipated power under an applied current caused Joule-heating of the carbon nanotube and its sur-

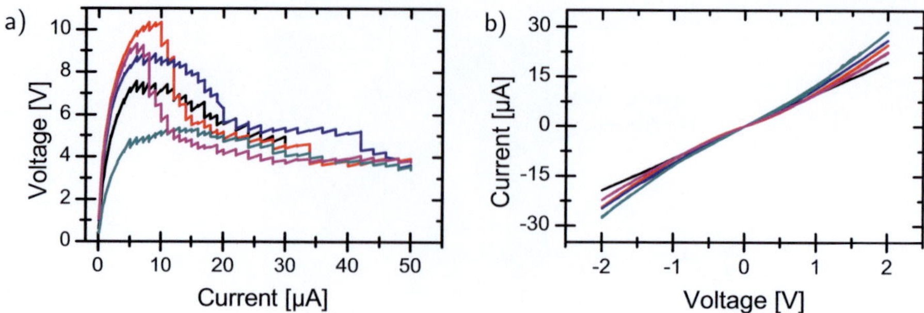

Figure 4.7: a) Conditioning of typical thin-film devices. The current driven through the device was increased stepwise up to the pre-set value of 50 µA. The voltage was continuously monitored. The resistance decreased permanently, as can be seen in the finally recorded I-V curves in b).

roundings. Thermally activated effects then led to a permanent increase in conductance of the device:

- The CNTs used in this work are wrapped in an organic surfactant, making them soluble in water. These surfactants do not withstand high temperatures, and either desorb from the nanotube or decompose into their constituents. The electronic properties of surfactant-wrapped CNTs are often perturbed compared to pristine, clean CNTs.
- High temperatures also allow curing of defects in the lattice of the CNT itself, again improving its electronic properties. This self-healing has been observed in molecular-dynamics simulations [163] as well as in transmission electron microscope studies [164].

A typical current-voltage trace of five thin-film devices during conditioning is plotted in Fig. 4.7a. The current was increased stepwise up to a value of 50 µA and the voltage monitored simultaneously. The resistance started to drop when a current of $\approx$5 µA was sourced through the device. In Fig. 4.7b the final I-V curves are plotted. After conditioning, the devices showed ohmic behavior, as expected for metallic carbon nanotubes.

4.1.4 Results of Nanogap Fabrication

During the EBIO process a constant source-drain voltage V_{SD} of maximum 1 V was applied and the current continuously measured in intervals of 50 ms. To cut the carbon nanotubes, an electron-beam line scan across the nanotube was performed while injecting oxygen gas into the microscope chamber. During line scans the microscope magnification was adjusted to either 25 kX or 50 kX, yielding a line scan width

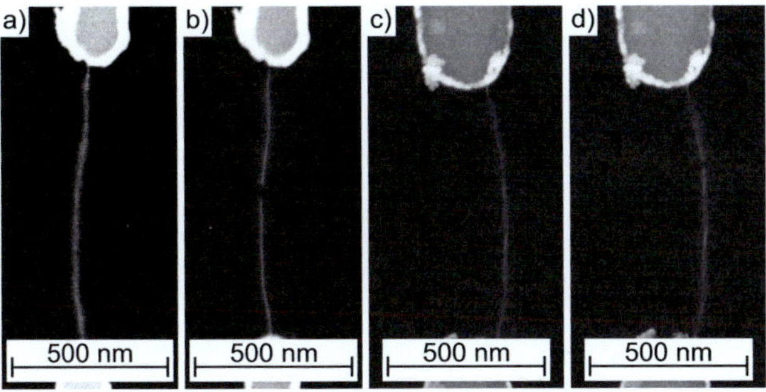

Figure 4.8: a,b) Secondary electron (SE) images of a CNT on a silicon oxide substrate before (a) and after (b) gap formation, recorded with a backgate voltage of +10 V. c,d) SE images of a CNT on a silicon nitride membrane before (c) and after (d) gap formation.

of 4.57 µm and 2.29 µm, respectively. The primary electron-beam current used was ≈ 100 pA, which yields a line dose of ≈ 21.8 µC/m and ≈ 43.7 µC/m per second, respectively. The Acceleration voltage of the primary electrons (PEs) was set to 10 kV. Scale-calibrated images were used to assess the reproducibility of gap sizes.

For SEM images of single-tube devices on silicon oxide and free-standing silicon nitride membranes before and after gap formation, see Fig. 4.8. The gaps shown are ≈ 20 nm wide. The average gap size over 62 devices on silicon oxide substrates is (19 ± 5) nm. For a histogram of gap sizes see Fig. 4.9.

The formation of a gap has also been confirmed by topographic measurements with an atomic force microscope. EBIO is not limited to the cutting of individual nanotubes, but can also be extended to form gaps in devices with multiple nanotubes, see Fig. 4.10 for an example. Nanogaps of a similar size can be produced at the same relative position in all of the nanotubes in parallel. The formation of such a line of nanogaps in thin-film devices is demonstrated in Fig. 4.11. Such a result cannot be obtained by current-induced oxidation [165, 166].

The critical line dose ξ_C required for gap formation has been determined from in-situ conductance measurements. Fig. 4.12 shows the typical conductance reduction of a mSWNT-device on SiO_2/Si and Si_3N_4, during the EBIO process. The traces of the conductance G have been normalized to their initial values G_i, at the beginning of the EBIO process at $t = 0$. G_i is typically on the order of $(40\,k\Omega)^{-1}$ after current-annealing. G decreases with ξ roughly exponentially by two orders of magnitude, before a sudden drop by additional 3-4 orders of magnitude down to the electron-beam-induced residual conductance of less than $(10\,G\Omega)^{-1}$ is recorded. This last step indicates the formation of a gap in the nanotube. There are no plateaus in G which could indicate a stepwise introduction of defects. Typically, ξ_C is on the order of a few mC/m.

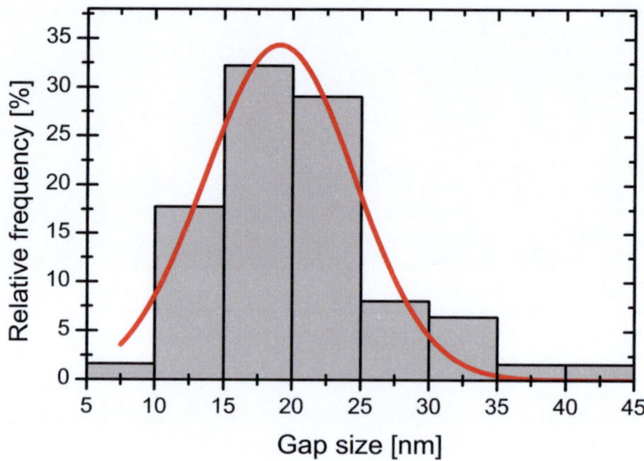

Figure 4.9: Histogram of nanogap sizes over 62 devices on silicon oxide substrates, fabricated by electron-beam-induced oxidation. The average gap size is (19 ± 5) nm.

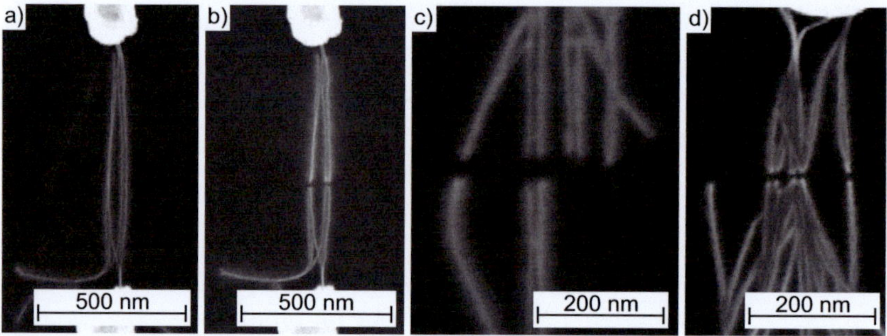

Figure 4.10: Secondary electron images of few-CNT-devices on a silicon oxide substrate before (a) and after (b,c,d) gap formation, recorded with a backgate voltage of +10 V.

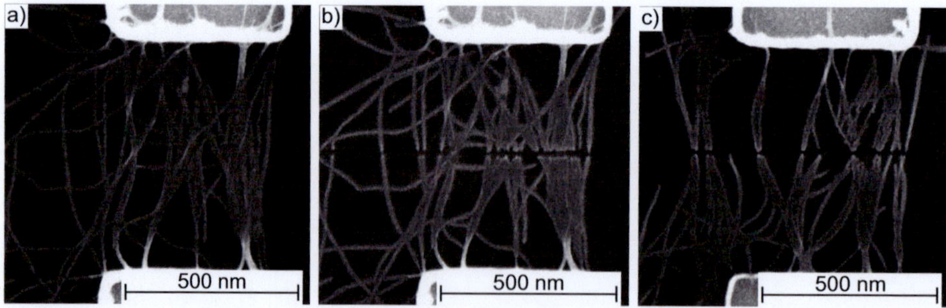

Figure 4.11: Secondary electron images of CNT thin-film devices before (a) and after (b,c) gap formation, recorded with a backgate voltage of +10 V.

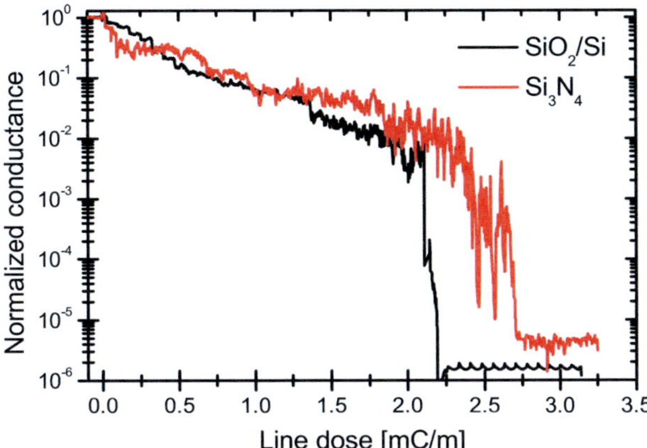

Figure 4.12: Normalized conductance G/G_i vs. line dose ξ recorded during the EBIO process for a single-tube device on SiO_2/Si and on a Si_3N_4 membrane.

Conductance traces for several thin-film devices during the EBIO process are shown in Fig. 4.13. The critical dose is comparable to or larger than the dose for single-tube devices, which is probably due to the presence of overlapping nanotubes in some of the thin-film devices.

4.1.5 Discussion and Conclusion

The conductance drop remained irreversible even at $V_\mathrm{SD} \gg 10\,\mathrm{V}$. This behavior is in marked contrast to the continuous electron-beam-induced metal-to-insulator transitions which are caused by charges trapped in the substrate surface [167]. It has been shown that such transitions require a line dose of $\approx$200 µC/m at 10 kV primary-electron energy. During the EBIO process, such a small dose causes only a small reduction in G. We assume that charging is not important here since oxygen ions are compensating electron-beam-induced surface charges.

In collaboration with Dr. Carl Georg Frase and Klaus-Peter Johnsen at the Physikalisch-Technische Bundesanstalt (PTB) in Braunschweig, Monte Carlo simulations of secondary electron generation in SiO_2/Si substrates and free-standing Si_3N_4 membranes were conducted using the Monte Carlo SEM code [168] developed at the PTB. The simulation comprised one million primary-beam electrons impinging on either substrate with a kinetic energy of 10 keV, with a lateral spread of 2 nm (FWHM). Top-view plots of the number of escaping secondary and backscattered electrons from both substrates are seen in Fig. 4.14. The radial distribution of SE2 and BSE electrons is much wider for the SiO_2/Si substrate compared to the free-standing membrane, and

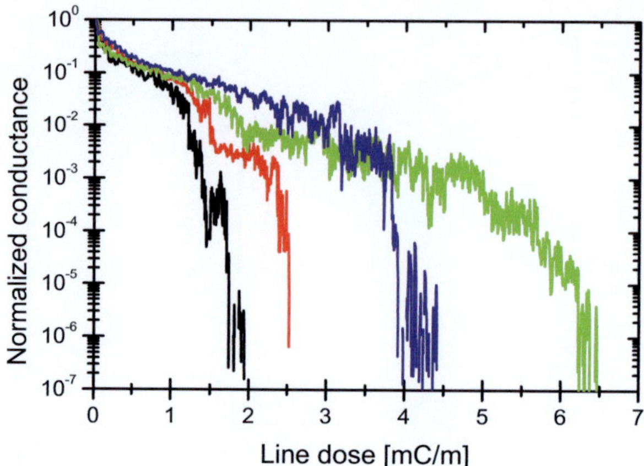

Figure 4.13: Normalized conductance G/G_i vs. line dose ξ recorded during the EBIO process for several thin-film devices on SiO_2/Si.

their total number is higher. This is expected, as the thick oxide as well as the substrate underneath lead to increased backscattering. However, the number of generated SE1 electrons at the irradiation spot is similar for both substrates, on the order of 20000. The lack of a significant difference between EBIO on a thick SiO_2/Si substrate and a free-standing Si_3N_4 membrane implies that backscattered electrons and BSE-generated secondary electrons (SE2) play a minor role.

We assume that the gap acquires its large width of $\approx 20\,\mathrm{nm}$ either because of an exothermic chain reaction where the gap size is determined by the oxygen partial pressure in the system [14], or due to the diffusion of oxygen radicals.

In conclusion, reliable and lithographically precise fabrication of mSWNT electrodes separated by a small gap generated by electron-beam-induced oxidation was demonstrated for individual CNT and CNT thin-film devices. Nanogaps can be fabricated reliably down to a size of $(19 \pm 5)\,\mathrm{nm}$. The independence of the gap size of the choice of substrate suggests that the process is driven by primary and SE1 electrons. The gaps produced by EBIO are still large compared to molecular scales. However, they have an advantageous size e.g. for addressing phase-change materials where reproducible gap sizes are required to obtain reproducible switching properties [15].

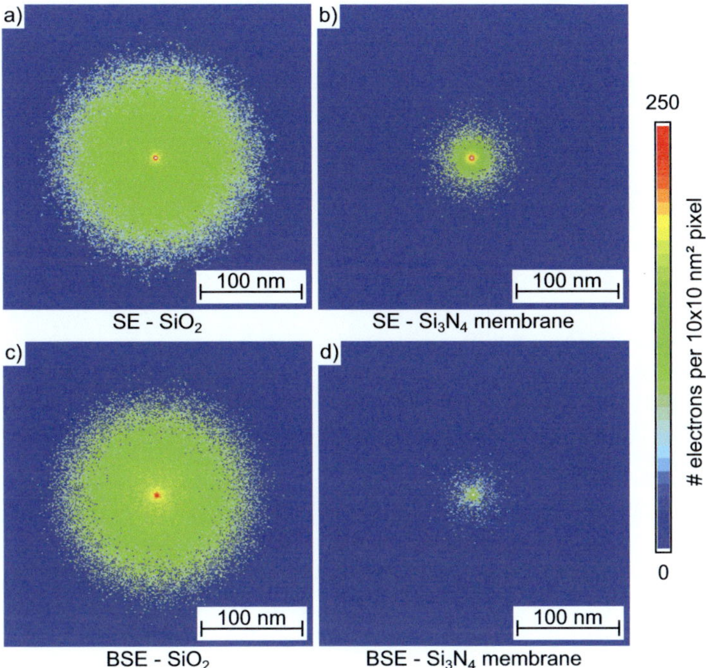

Figure 4.14: Monte Carlo simulations of 10 keV primary-beam electrons impinging on a,c) 800 nm SiO_2 on a silicon substrate and b,d) 50 nm free-standing Si_3N_4 membranes. Plots represent top views on substrates and indicate the number of escaping secondary electrons (SEs) and backscattered electrons (BSEs) per $10 \times 10\,\text{nm}^2$ pixel. A central spot of SE emission at the irradiation spot has been omitted from both SE plots, in both cases its value is ≈ 20000.

4.2 Nanogaps by Helium-Ion-Beam Lithography

In this section, the fabrication of nanogaps in carbon nanotubes by helium-ion sputtering is presented. The experiments were conducted in collaboration with Henning Vieker and Dr. André Beyer of Prof. Armin Gölzhäuser's group at the Faculty of Physics at Bielefeld University. Some of these results on the nanogap fabrication and the first results of electrical measurements of OPE-A molecules placed into these nanogaps have been published in [1].

4.2.1 Device Fabrication

Electrodes were fabricated on degenerately doped silicon substrates with 800 nm thermal oxide. A two-layer resist system was used to decrease sidewalling effects and to improve the lift-off. Two different resist formulations of poly(methyl methacrylate) (PMMA) were used. The bottom layer was PMMA 600K EL11 (600 kDa avg. weight, 11 % solid content in ethyl lactate). After spin-coating at 6000 rpm, which led to a thickness of $\approx$ 180 nm, the resist was baked in an oven at 165 °C for 30 min. Subsequently, a layer of PMMA 950K A4.5 was spin-coated at 6000 rpm and again baked at 165 °C for 30 min. Standard electron-beam lithography was used to write the electrode patterns into the resist, which was then developed with a 1:3 mixture of methyl isobutyl ketone (MIBK) in isopropanol for 30 seconds.

To achieve an almost flat surface after metal deposition, reactive ion etching was used to etch into the silicon oxide prior to metal deposition. A CHF_3 plasma with a small addition of O_2 and Ar gas at a pressure of 50 mTorr was used for this purpose. Each etching cycle consisted of 30 seconds at 100 W power; afterwards the chamber was flushed with Ar and oxygen gas and pumped down to 10^{-5} mTorr. After five cycles, the trench in the oxide was $\approx$ 50 nm deep. Tungsten was then sputter-deposited. After 90 seconds, the amount of metal had approximately filled the trench and sputtering was stopped. Lift-off of the mask and any undesired metal on top was then performed in an acetone bath with ultrasonication.

Suspensions of metallic single-walled carbon nanotubes (mSWNTs) were prepared by Dr. Benjamin S. Flavel by the method described in Sec. 2.4.2. mSWNTs with a diameter distribution of (1.2 ± 0.2) nm were used in this work. For the deposition of CNTs between the metallic contacts, dielectrophoresis was used as explained in Sec. 2.4.3. An alternating voltage with a frequency of 300 kHz and a peak-to-peak voltage between 1.0 and 1.3 V was applied between source and drain contacts, while a $\approx$ 50-µl drop of CNT dispersion with a concentration of $\approx$ 5 CNTs per µm^3 was placed on the device. After 5 minutes, the drop was first diluted with bidistilled water, followed by methanol and finally allowed to dry. CNT deposition was assessed with a conventional SEM. Different electrode geometries were used, see Fig. 4.16 for SEM images after tube deposition. For sharp electrode tips no significant difference

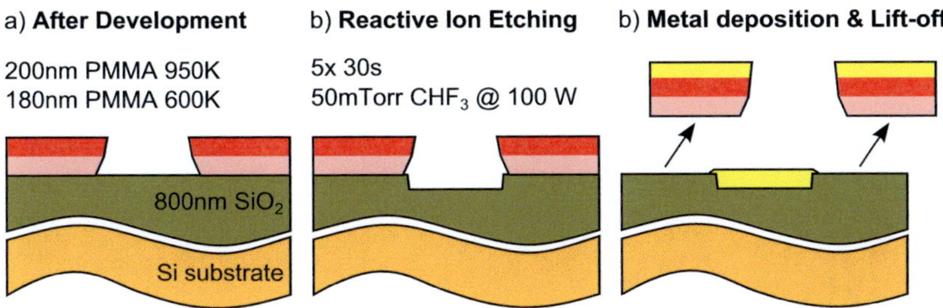

Figure 4.15: Fabrication of "buried" metallic electrodes. a) The bottom photoresist layer is more sensitive than the top resist, leading to an undercut profile after development. b) Reactive ion etching is used to transfer the electrode pattern into the oxide layer. c) Tungsten is sputter-deposited into the trench and the photoresist with undesired metal on top is removed in an acetone bath.

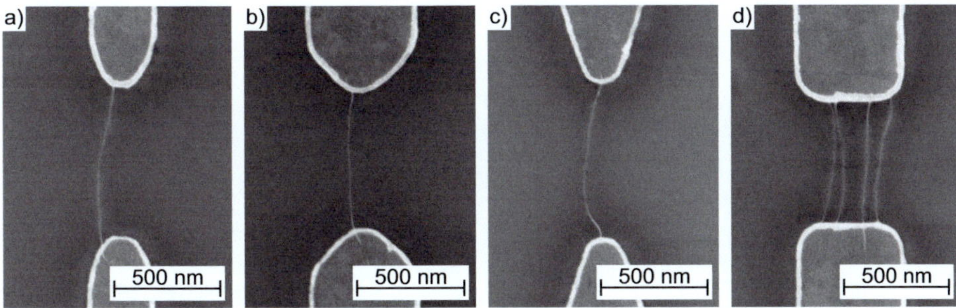

Figure 4.16: a-d) SEM images of devices with different electrode geometries after dielectrophoresis of CNTs. For sharp electrode tips no significant difference in the number of carbon nanotubes per contact was found, whereas broader electrodes showed a tendency towards side-by-side deposition of multiple tubes.

in the number of carbon nanotubes per contact was found, whereas broader electrodes showed a tendency towards side-by-side deposition of multiple tubes. The number of tubes per contact was counted and is plotted in a histogram vs. deposition voltage in Fig. 4.17. As expected, there is a tendency towards fewer carbon nanotubes per contact for lower deposition voltages. Before the samples were transferred to the HIM, they were annealed in a vacuum oven ($p = 10^{-6}$ mbar) at 600 °C for 30 minutes to eliminate any effects from electron-beam exposure in the SEM.

As explained in Sec. 2.6, the HIM used in this work allows imaging similar to an SEM. For all images, the secondary-electron detector was employed. The acceleration voltage of the HIM was always set to 35 kV and the 5-μm aperture was used, resulting in a spot size of below 1 nm and a beam current of ≈ 0.4 pA. A comparison of conventional electron microscopy and the HIM can be seen in Fig. 4.18. In the HIM image, even bundles consisting of only a few tubes can be distinguished from single tubes.

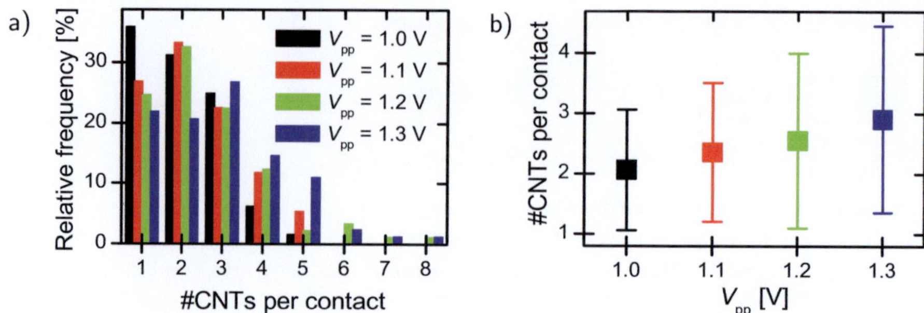

Figure 4.17: Statistics of CNT deposition using a range of peak-to-peak voltages from 1.0 V to 1.3 V. As expected, there is a trend towards fewer CNTs per device when using a smaller deposition voltage.

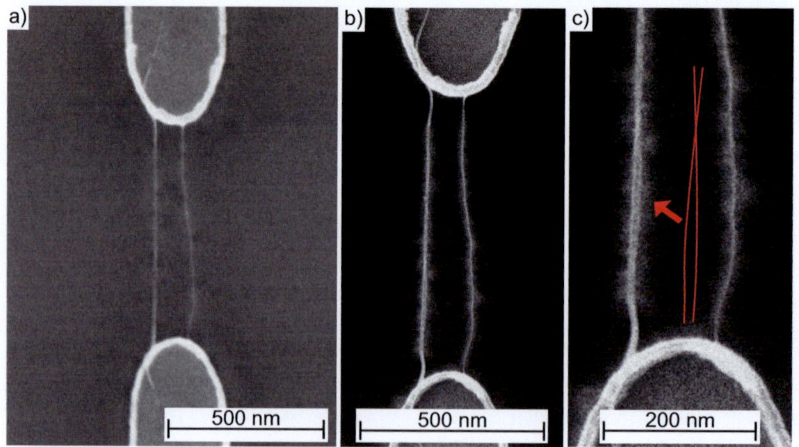

Figure 4.18: Comparsion of SEM and HIM images of the same device. a) SEM image after tube deposition. b) HIM image of the same device. c) HIM zoom. The bundle of tubes on the left can be seen to twist in the helium ion image, as indicated.

The signal-to-noise ratio of an image and the implanted ion dose depend on the beam current, the dwell time per pixel and the averaging settings. To minimize the ion dose, fast alignment images before sputtering were recorded using the following settings: Pixel spacing 1 nm, dwell time 0.5 µs, no averaging. This led to a line dose of 0.2 nC/m per scan line. Later, slow scans for characterization were performed with fixed parameters: A pixel spacing of 5 Å, a dwell time of 0.5 µs and 32 x line averaging. This led to a line dose of 13 nC/m per scan line.

4.2.2 Sputtering of Nanogaps

Helium-ion sputtering as explained in Sec. 2.6.2 was used to fabricate nanogaps in the carbon nanotubes. To reduce hydrocarbon deposition on the surface, all samples were stored in the HIM chamber under high vacuum for at least several hours. The chamber pressure typically reached $2.5 \cdot 10^{-7}$ mbar before experiments were started. In order to cut CNTs, a single pixel line with a pixel spacing of 2.5 Å and a dwell time in the millisecond range was scanned across a CNT. To align this line perpendicularly to the nanotube, a fast scan was performed before the lithography was started.

To determine the critical dose for gap formation, voltage-contrast microscopy was employed. This was realized by grounding one of the two metal electrodes. The other electrode remained floating, albeit connected to the grounded electrode by the mSWNT. After a single pixel line was scanned across the nanotube, a slow scan image of the device was acquired. Once an electrically insulating nanogap was formed in the metallic CNT, the floating electrode accumulated positive charges, thereby inhibiting secondary electrons from reaching the detector. Consequently the floating electrode appeared darker in the image, while the grounded electrode and the CNT segment connected to it appeared brighter, see Fig. 4.19 for examples. The dwell time per pixel was increased until a different contrast of the two electrodes was observed. Using this experimental procedure, the critical dose for gap formation was determined to be $\approx 24\,\mu$C/m. Before proceeding to cut further nanogaps, the beam current was measured and the pixel dwell time adjusted accordingly to accommodate this value. The line doses implanted by fast and slow scan images are at least three orders of magnitude lower, and thus have a negligible sputtering effect.

4.2.3 Characterization of Nanogaps

Helium-Ion Microscopy

In order to precisely measure the size of the gap formed, the electrostatic charging of floating electrodes had to be avoided as charging always causes drifts. Furthermore, the different secondary-electron intensities on opposite sides of the gap would make

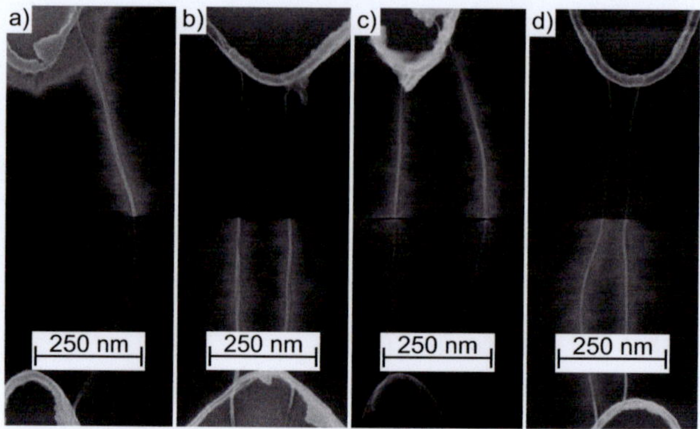

Figure 4.19: Voltage-contrast effect on different devices. Prior to these images, an insulating gap was formed in the center of the carbon nanotube. One electrode and the attached CNT segment is electrically grounded and appears brighter, while the other is electrically floating and charges up positively, thus appearing darker.

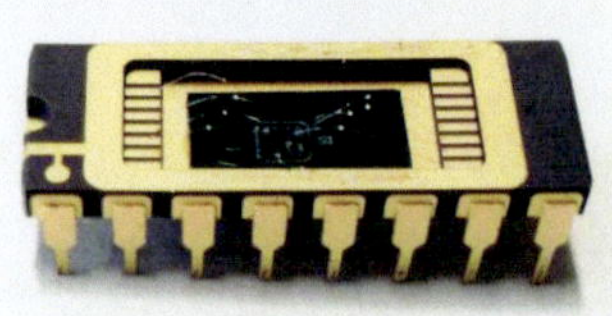

Figure 4.20: 16-pin dual in-line package chip carrier with a mounted sample. Contact pads of the common drain electrode, the backgate and source electrodes of the nanogap devices were wire-bonded with Al wires to pins on the chip carrier.

an analysis difficult. To avoid charging, the samples were mounted in commercial 16-pin dual in-line package chip carriers and the common drain electrode and the source electrodes of devices were wire-bonded to pins on the chip carrier with Al wires, see Fig. 4.20 for a photograph. A customized sample holder in the HIM connected all pins to the stage/ground potential. After a fast alignment scan, shown in Fig. 4.21b, a nanogap was fabricated and a detailed slow scan image acquired, see Fig. 4.21c/d.

Secondary-electron-intensity profiles were then recorded across 14 nanogaps in different carbon nanotube devices. The resulting curves were fitted with inverted Gaussians, see Fig. 4.22. Three profiles including their Gaussian fit as well as a scatter plot of profiles from all devices are plotted. The profiles are remarkably similar to each other. A histogram of FWHM from Gaussian fits of the intensity profiles is plotted in Fig. 4.23. With an average nanogap size of (2.8 ± 0.6) nm, direct helium-ion sputtering is more precise by almost an order of magnitude than electron-beam-induced etching and a factor of 2-3 better than the smallest gaps achievable by current-induced breakdown. Most striking is the comparably narrow gap size distribution, which is an

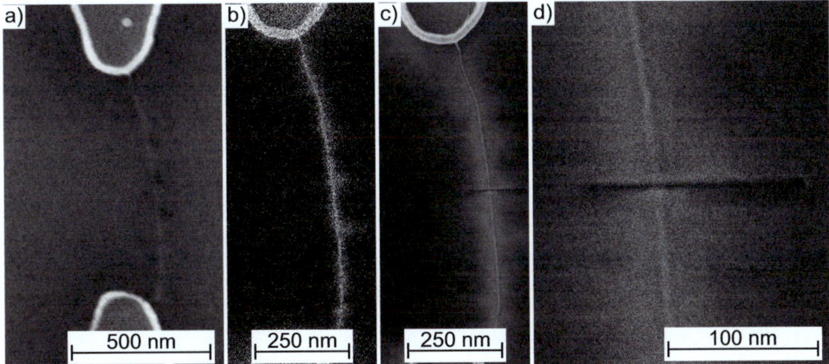

Figure 4.21: a) SEM image of a CNT device after deposition. b) Fast scan of the same device in the helium-ion microscope before cutting. c) Slow scan after cutting. d) Zoom on to the area of the cut.

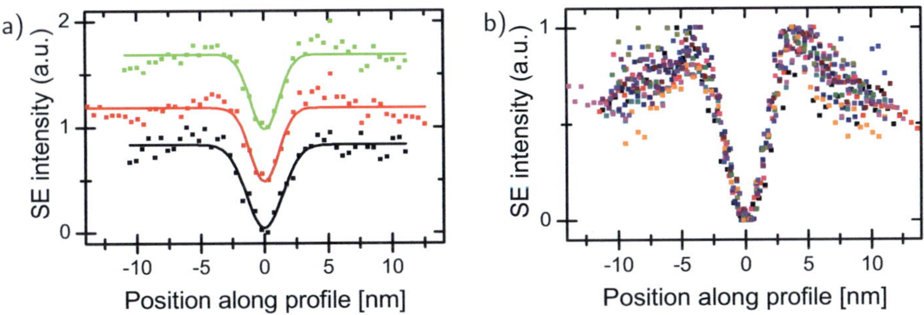

Figure 4.22: a) Stacked profiles of three devices. b) Scatter plot of profiles of all devices.

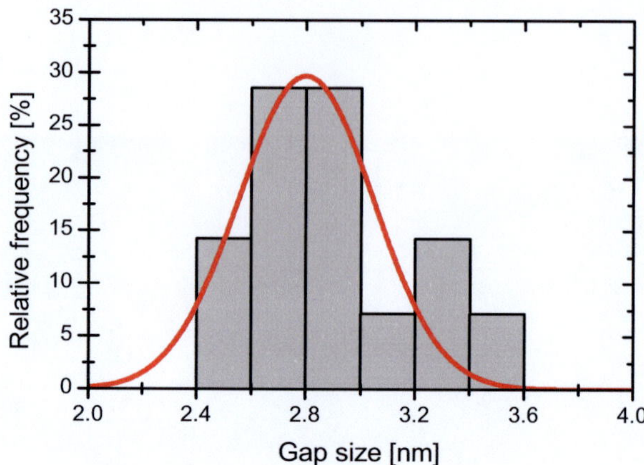

Figure 4.23: Histogram of nanogap sizes over 14 devices, fabricated by helium-ion sputtering. The average nanogap size is (2.8 ± 0.6) nm.

indication of the highly reproducible nature of this method. As we were targeting the smallest possible gap size for the subsequent insertion of molecules we did not explore the formation of larger gaps. Further detailed images of nanogaps in single-CNT and few-CNT devices are shown in Fig. 4.24.

As a side note, the fabrication of a parallel line of nanogaps across CNT thin-film devices, as demonstrated by electron-beam-induced oxidation in Sec. 4.1.4, is achievable by helium-ion sputtering as well, see Fig. 4.25. However, as our interest lies in the fabrication of single electrode pairs for molecular electronics, we did not pursue the fabrication of nanogaps in thin-film devices.

Atomic Force Microscopy

Atomic force microscopy (AFM) measurements were attempted on all nanogap devices using a Veeco MultiMode AFM in tapping mode with commercially available 15 nm-radius silicon nitride tips. However, the lateral resolution was typically not sufficient to resolve the nanogap, showing an indentation with a FWHM from 20-30 nm, except for the measurement shown in Fig. 4.26b. The HIM image in Fig. 4.26a is the alignment image used before nanogap sputtering and shows the intended nanogap position. At the same position a trench is observed in the AFM height image. A detailed slow scan image was not acquired with the HIM after sputtering, because this particular device was intended for molecular contacting experiments. A profile across the nanogap along the nanotube can be fitted with an inverted Gaussian and shows a FWHM of the nanogap of ≈ 3.5 nm, as indicated in Fig. 4.26. The depth of the trench is ≈ 2 nm. Along the sputtering line scan, a trench is visible in the substrate as well.

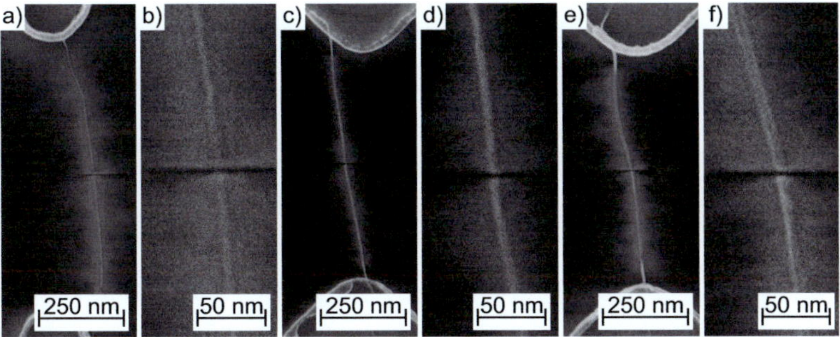

Figure 4.24: Detailed images of nanogaps fabricated by helium-ion sputtering. a)-d) Overview image and corresponding zoom on nanogaps in single-CNT devices. e,f) Nanogap in a bundle of two CNTs.

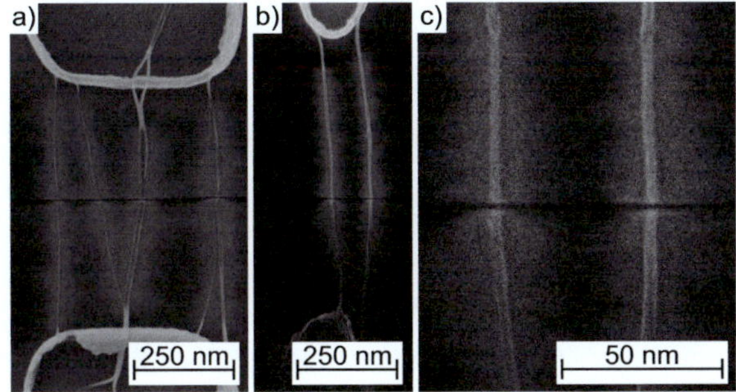

Figure 4.25: Parallel fabrication of nanogaps in thin-film and few-CNT devices by helium-ion sputtering.

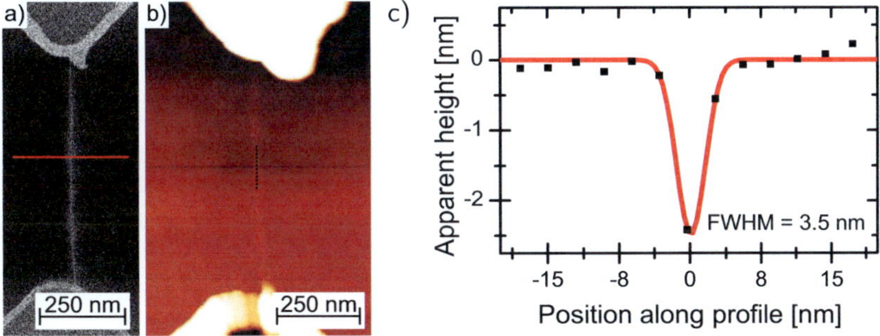

Figure 4.26: Atomic force microscope characterization. a) HIM image of device before nanogap fabrication, indicating the desired position of the nanogap. b) AFM topography of the same device, showing an indentation at that position. c) Height profile extracted from the AFM data.

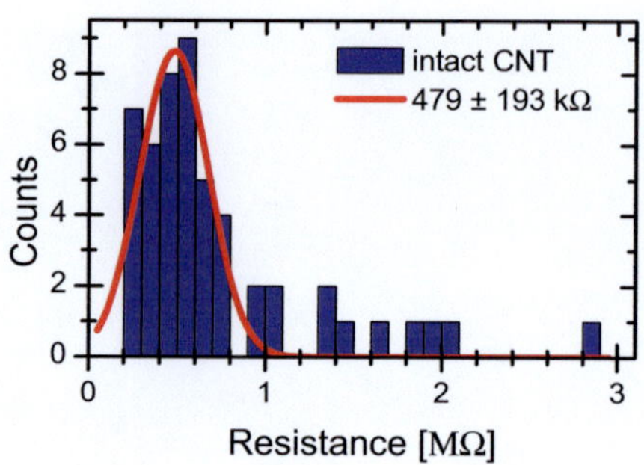

Figure 4.27: Conductance histogram of pristine CNT devices.

Electrical Characterization

Electrical characterization of pristine CNT devices showed an ohmic current-voltage behavior with a typical low-bias resistance of $(479 \pm 193)\,\mathrm{k\Omega}$, see Fig. 4.27 for a conductance histogram. This is comparable to 1.2 nm diameter mSWNTs on Pd electrodes [169].

Devices that were electrically characterized or used for further experiments were not imaged with a detailed scan after gap formation to avoid any detrimental effects of the ion beam such as unwanted sputter damage or ion or charge implantation. To measure currents down to the femto-Ampère range, an atmospheric probe station was outfitted with probe arms with triax connectors. Three source/measure units of an Agilent 4155C Semiconductor Parameter Analyzer were then used for guarded measurements of nanogap devices at room temperature. A reference measurement of the leakage current through the prober insulation is shown in Fig. 4.28. A parasitic resistance of $1.36\,\mathrm{P\Omega}$ was extracted for the measurement setup.

mSWNTs with nanogaps had a resistance of $(643 \pm 311)\,\mathrm{T\Omega}$, which is nine orders of magnitude higher than that of pristine devices. A typical I-V curve and a conductance histogram over all devices is plotted in Fig. 4.29. Some of the obtained resistance values for nanogaps are of the same order of magnitude than the parasitic resistance of our setup. Diffuse charge transport through the air gap or through trap states on the substrate surface may be the cause of the observed current. We attempted to measure tunneling or field-emission currents through the air gap and to correlate the current with the HIM derived gap size. However, despite the large electric fields of up to 5 V/nm in these nanogaps, we were not able to detect any sign of field emission. With the Simmons model the expected tunneling current through a vacuum barrier can be estimated [170]. Assuming a junction area of $1\,\mathrm{nm}^2$, corresponding to the cross-sectional area of one carbon nanotube and a barrier height of 5 eV, corresponding to the work function of CNTs [171], a tunneling current on the order of 10^{-20} A is expected for a vacuum gap size of 2 nm and an applied voltage of 5 V, well below

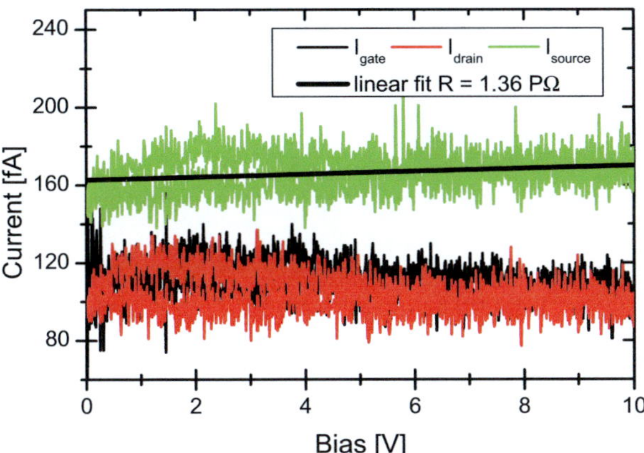

Figure 4.28: Measurement of parasitic resistance of the measurement setup. Three source/measure units (SMUs) of an Agilent 4155C Semiconductor Parameter Analyzer are connected to a probe arm with a triax cable each. A voltage from from 0 to 10 V and backwards was swept from the I_{source}-SMU with the probers separated by a macroscopic air gap. The offset current at zero voltage is due to the imperfect zero offset adjustment of the measurement setup.

the detection limit of our measurement setup. Furthermore, the expected tunneling current decreases by an order of magnitude with every Ångström in gap size.

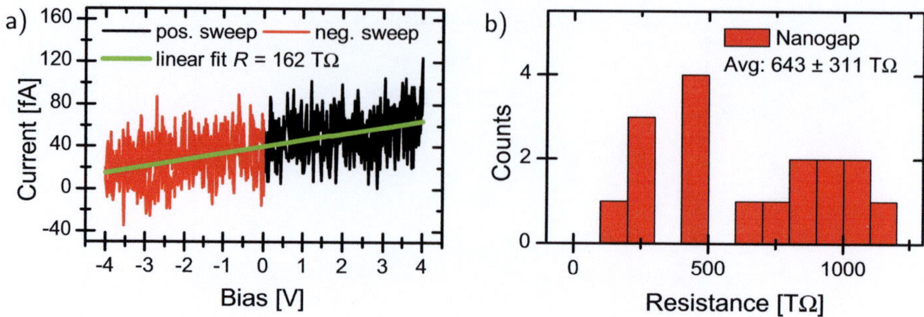

Figure 4.29: a) *I-V* characteristic of a nanogap device. The offset current at zero voltage is due to the imperfect zero offset cancellation of the measurement setup. b) Resistance histogram of nanogap devices fabricated by helium-ion sputtering.

4.2.4 Estimation of Cross-Section

To estimate the cross-section of helium-ion sputtering of carbon nanotubes, a simplified model of a suspended carbon nanotube is considered, excluding any recoil effects from surface atoms. Metallic carbon nanotubes with a diameter distribution of (1 ± 0.2) nm were used for the experiments. However, the exact distribution of chiralities of the nanotubes is unknown. For armchair and zig-zag nanotubes, the number of carbon atoms per length can be easily estimated:

- A (n,n) armchair nanotube comprises n armchairs along its circumference, with three carbon atoms each. They are spaced 2.45 Å apart along the length of the nanotube. All armchair nanotubes are metallic.
- A $(n,0)$ zig-zag nanotube comprises n zig-zags along its circumference, with two carbon atoms each. Two rows of zig-zags per $(2.83 + 1.42)$ Å constitute the nanotube. Zig-zag nanotubes with $n = 3l$ $(l \in \mathbb{N}^*)$ are metallic.

The number of carbon atoms per nanometer was thus estimated for various metallic armchair and zig-zag nanotubes with a diameter in the range of (1.2 ± 0.2) nm, see Table 4.1. As expected, the number of carbon atoms per nanometer is correlated with the diameter of the nanotube. The values for other, chiral metallic carbon nanotubes are thus in the same range. The average number of carbon atoms per nanometer over the listed chiralities is 150 ± 16.

The size of the nanogap in our CNTs was determined from secondary-electron intensity profiles to (2.8 ± 0.6) nm, as shown in Fig. 4.23. Consequently, the number of carbon atoms sputtered from the nanogap was $\approx 421 \pm 135$. The helium-ion line dose for gap formation was determined to $\approx 24\,\mu$C/m, see Sec. 4.2.2. This corresponds to $1.5 \cdot 10^5$ ions/nm. The ion beam had a FWHM of less than 1 nm and was scanned perpendicularly across the nanotubes. Due to their average diameter of 1 nm, the

Chirality	**Diameter**	**Carbon atoms/nm**
(8,8)	1.08 nm	130.6
(9,9)	1.22 nm	146.9
(10,10)	1.36 nm	163.3
(15,0)	1.17 nm	141.2
(18,0)	1.41 nm	169.4

Table 4.1: Estimation of the number of carbon atoms per nanometer for various metallic armchair and zig-zag nanotubes with a diameter in the range of (1.2 ± 0.2) nm.

number of impacting ions can be estimated to $1.5 \cdot 10^5$. So, for every 356 ± 114 ions, one carbon atom is ejected from a nanotube. Lehtinen et al. used a Monte Carlo method to simulate ion impacts onto a suspended graphene sheet [139]. For helium ions with a kinetic energy of 30 keV, their simulation yields one sputtered carbon atom per 416 ions[1]. The cross-section for helium-ion sputtering of carbon nanotubes is hence similar to that for sputtering of graphene.

[1]This value is derived from Fig. 3e in [139]. The graphene target consisted of 800 atoms, and 1.9 % of the atoms were sputtered after an ion dose of 10^{-15} C. This implied that 6241 ions sputtered 15 carbon atoms.

5 Direct Contacting of OPE-A

5.1 Deposition of OPE-A into Nanogaps

As it was possible to fabricate nanogaps by helium-ion sputtering with a size where one OPE-A molecule should be able to bridge the gap, it was attempted to deposit OPE-A molecules into the nanogaps for direct electrical measurements on it. Similar to the preparation of Au(111) surfaces for scanning tunneling microscope experiments, solutions of OPE-A molecules were prepared in ultra-clean dichloromethane with molecular concentrations of $\approx 1\,\mu g/ml$.

For the deposition, a probe voltage of typically 1 V was applied across a nanogap and the source-drain current monitored. To test the effect of the solvent without the molecule, a drop of DCM with a volume of $\approx 20\,\mu l$ was placed on the device and allowed to evaporate. The corresponding trace of current vs. time is shown in Fig. 5.1a. We observe that during the presence of the solvent drop on the device the current is rather high. We attribute this to the residual ionic conductivity of the solvent, e.g. caused by traces of salts from chemical drying. Once the drop has evaporated after roughly two minutes, the resistance of the device returns to the level of a pristine nanogap. If, however, solvent with the OPE-A molecule is used for the same experiment, the resistance after deposition is reproducibly lower after the drop has evaporated, as shown in Fig. 5.1b. This lower-resistance state is indicative of molecule deposition across the nanogap..

Molecule deposition can be repeatedly performed on a single device, as it is possible to wash off the molecule with clean solvent. This is demonstrated in Fig. 5.2a. A nanogap device with molecules in the gap is subjected to repeated cycles of clean solvent and blow-drying. Already after the first rinsing the resistance can be seen to decrease to the value of the nanogap. After rinsing, when the solution with molecules is used again, the current stabilizes at a higher value after drop evaporation.

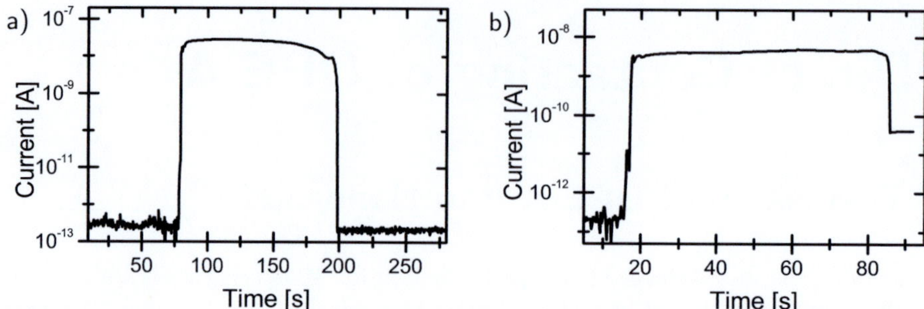

Figure 5.1: a) A probe voltage of 1 V is applied across the nanogap and a drop of dichloromethane (DCM) placed on it. The current level increases due to ionic impurities in the solvent. When the drop has evaporated, the current level returns to the nanogap level. b) Repeating the procedure with a drop of OPE-A dissolved in DCM at $\approx 1\,\mu g/\mu l$, the current does not return to the nanogap level and instead stabilizes at a higher value.

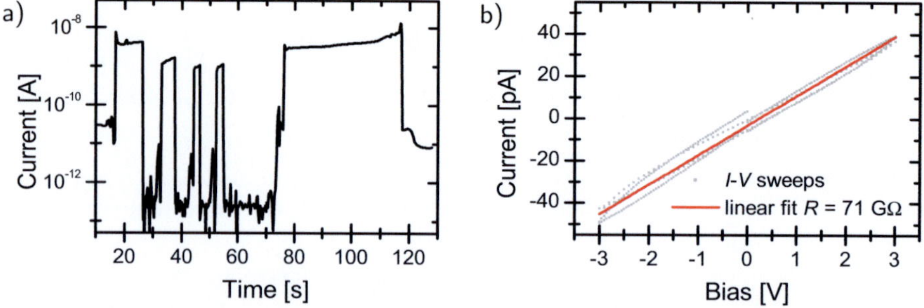

Figure 5.2: a) Washing off and re-deposition of OPE-A molecules. A device with molecules in the nanogap is rinsed with clean dichloromethane four times. The current returns to the nanogap level already after the first rinsing. Then, a solution of OPE-A molecules in DCM is used and the current stabilizes again at a higher level after evaporation. b) Typical I-V-curve recorded with OPE-A molecules in the nanogap. The curve is linear over a wide voltage range.

5.2 Electrical Characterization

A typical *I-V* curve of a nanogap with OPE-A molecules is plotted in Fig. 5.1b. The *I-V* curve is ohmic, with no apparent features. Molecule deposition was carried out in total for 14 devices on four different chips. It was typically repeated four times for a single device, resulting in 50 successful molecule depositions out of 54 attempts. A resistance histogram of all devices with a molecule in the gap is plotted in Fig. 5.3a. The average resistance over all of these devices is $(90 \pm 85)\,\mathrm{G}\Omega$. In Fig. 5.3b, the histogram includes only devices were the molecule deposition was repeated at least twice. The result is color-coded per device. No correlation of the resistance value with a molecule in the nanogap is observed with the specific device number.

A combined histogram of resistances measured during the fabrication process is plotted in Fig. 5.4. It spans ten orders of magnitude, from intact carbon nanotube devices to carbon nanotubes with a nanogap, with the resistance values of devices after molecule deposition at its center.

Scanning tunneling microscope junction measurements on an OPE of similar length revealed a resistance of $\approx 200\,\mathrm{M}\Omega$ for a single molecule, albeit with the molecule covalently bonded to gold on both sides [25]. By trying to detach an OPE-A molecule from the Au(111) surface with the STM tip and directly measuring its resistance, a similar experiment was attempted, see Sec. 3.3.2. At a lifting height of $\approx 3.8\,\text{Å}$, a resistance of $\approx 50\,\mathrm{M}\Omega$ was observed, which likely corresponded to a situation where only a small part of the molecule was detached from the surface. However, for the OPE-A molecule contacted by helium-ion sputtered nanogaps, a resistance almost three orders of magnitude higher is observed, with values spread over roughly two orders of magnitude.

Furthermore, *I-V* curves of nanogaps with deposited molecules show no discernible features. From fluorescence measurements (see Sec. 2.1.1), DFT calculations (see Table 3.1 in Sec. 3.7) and spectroscopic STM measurements presented in Sec. 3.4, the HOMO-LUMO gap of OPE-A molecules was determined to be within 3 eV to 4 eV. If the nanogap electrodes were bridged by a single molecule, this band gap should have been observable in the *I-V* curves. The observed ohmic current-voltage behavior can be explained by a diffusive charge transport mechanism, possibly across a film of OPE-A molecules lying on the nanogap and its surroundings.

The high variance of the observed resistance might be due to a varying imperfect attachment of the molecule's anchor group to the carbon nanotube, and the conformational freedom of the molecules. The backbone of the molecule was observed to be able to bend strongly in our STM study. DFT calculations of a free OPE molecule by Seminario et al. have shown that the HOMO and LUMO orbitals are no longer delocalized over the length of the backbone once a twist is introduced along its length [36]. Without an underlying surface to planarize the molecule in the CNT nanogap, it might assume an unfavorable conformation, greatly diminishing its conductance.

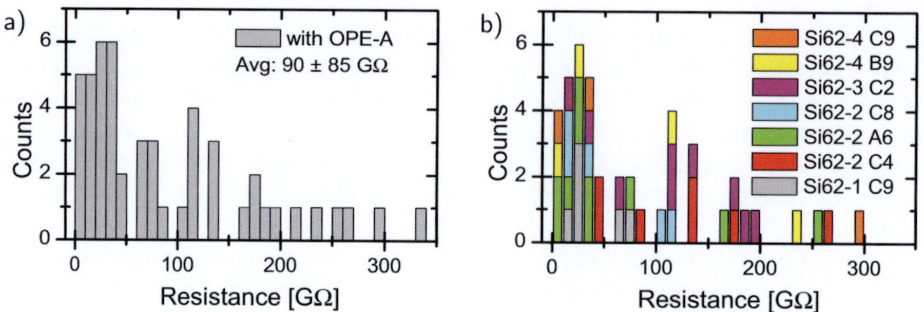

Figure 5.3: a) Resistance histogram over 50 devices after successful molecule deposition. b) The same histogram including only devices where the deposition process was repeated more than twice, color-coded per device. No correlation between a device and the observed resistance value with a molecule in the nanogap is observed.

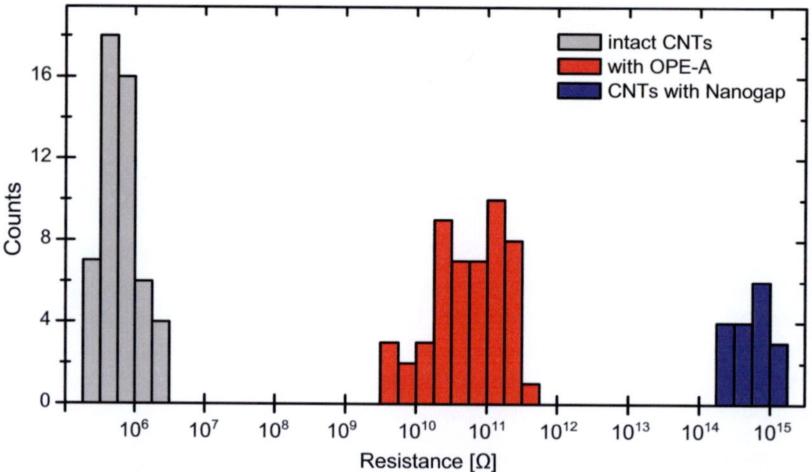

Figure 5.4: Combined resistance histograms covering the three states of a device, with four logarithmically-spaced bins per decade: (1 ; $\sqrt[4]{10} = 1.78$), (1.78 ; $\sqrt{10} = 3.16$), (3.16 ; $\sqrt[4]{10^3} = 5.62$), (5.62 ; 10). Pristine carbon nanotube devices exhibit an average resistance of 478 kΩ, devices with a nanogap of 643 TΩ. In between are devices where a molecule is bridging the nanogap, with an average resistance of 90 GΩ.

6 Conclusion and Outlook

In this work, the first scanning tunneling microscope (STM) observations of single oligo(phenylene ethynylene) (OPE) molecules were presented, consisting of four or five phenylene-ethynylene units and hexyl side chains. Both OPE variants, up to $\approx 5\,\text{nm}$ in length, were deposited on a Au(111) single-crystal surface and were observed to self-assemble at a sufficiently high concentration on the surface. Typically, the molecules arranged on the fcc region of the Au surface reconstruction. At low surface concentrations, isolated molecules were often grouped in pairs or triples, which strongly suggests they diffuse at temperatures up to $120\,^{\circ}\text{C}$ used for surface preparation. Through attractive and repulsive interactions with the STM tip, molecules could be separated from each other and moved around on the surface, even across the transition regions of the surface reconstruction. By approaching the anchor group of a molecule with the STM tip, the molecule would attach to the tip and it was possible to slightly lift it off the surface. However, molecules typically fell back to the surface at tip lift heights of maximum $5\,\text{Å}$. Still, measurements with attached molecules showed resistances of up to $\approx 50\,\text{M}\Omega$. As a large portion of the molecule was likely still in contact with the gold surface during this measurement, it is reasonable to assume that the resistance of a single molecule would be conceivably higher with the current flowing along the whole length of the backbone.

Despite the conductive surface, the highest occupied molecular orbital (HOMO) of both OPE variants could be directly imaged. However, the strong hybridization of Au surface states with unoccupied molecular orbitals inhibited a clear STM conductance map of the latter. The recorded shapes of the HOMO were confirmed by density functional theory calculations of the molecules' electronic structure. To include hybridization effects, one layer of an unreconstructed Au(111) surface was included in the DFT calculation for a shortened OPE molecule. The results confirmed that hybridization of the surface states is more relevant for the lowest unoccupied molecular orbitals.

The chemical structure of both molecules was directly imaged. By employing low bias voltages and an especially sharp, modified STM tip, a mode of contrast was achieved where individual carbon atoms and their bonds could be identified. Such a contrast is unusual for STM measurements and so far has only been reported by one other group in the literature. The observed contrast mode is still under discussion, and perhaps the results presented in this work will fuel further theoretical studies as to its origin. Finally, conformational switching in the alkyl chain substituents of a molecule were induced by the tunneling current and analyzed in detail. The analysis revealed that the switching is driven by single electrons. From the data threshold energies of $\approx 300\,\text{meV}$ were extracted, below which switching events were no longer observed.

To complement the STM characterization of the OPE molecules with direct electrical measurements in a device geometry, two fabrication methods of nanogaps into metallic carbon nanotubes were explored: Electron-beam-induced oxidation (EBIO) and helium-ion sputtering.

EBIO exploits the generation of reactive species from gas molecules introduced into the chamber of an electron microscope. At the focal point of the electron beam the primary-beam and secondary electrons thereby drive an oxidation reaction with a carbon nanotube. Devices were fabricated on thick silicon oxide substrates as well as on thin free-standing silicon nitride membranes to investigate the type of electrons responsible for etching. Metallic carbon nanotubes were deposited between palladium contacts by dielectrophoresis. By monitoring the conductance of the nanotube devices in-situ during the process of cutting, it was possible to extract the critical electron dose for gap formation, which is similar for both kinds of devices. With the help of Monte Carlo simulations of the secondary-electron generation, it was ascertained that both type-1 secondary and primary-beam electrons were responsible for the gap formation. Consequently, the observed gap size average of $\approx 20\,\text{nm}$ corresponded roughly to the secondary-electron escape depth of the substrates.

While the nanogaps produced by EBIO were still too large to be bridged by most single organic molecules, it was possible to fabricate them with a very high reproducibility. This could be advantageous for applications relying on a low gap size variance, such as the addressing of phase-change materials where the switching properties vary with the gap size [15]. Furthermore, the fabrication of parallel lines of nanogaps into thin-film devices was demonstrated, which is not possible, e.g., by current-induced oxidation [165].

The fabrication of nanogaps by helium-ion sputtering yielded nanogaps almost an order of magnitude smaller, with an average size of only $(2.8 \pm 0.6)\,\text{nm}$, sufficiently small to be bridged, e.g., by a single OPE molecule. A similar device layout as for EBIO, with carbon nanotubes contacted by metallic contacts, was used for these experiments. Whilst in-situ conductance measurements were not possible, voltage-contrast imaging was employed to determine when an insulating gap formed in a nanotube. However, high-resolution images could only be acquired of devices where charging effects were inhibited by grounding both sides of the nanotube. Electrical characterization of the nanogaps revealed a resistance nine orders of magnitude higher than that of pristine devices.

Finally, the first experiments towards directly contacting one of the OPE variants were presented. A probe voltage was applied on a nanogap device and solutions with and without the molecule in high-purity dichloromethane were allowed to dry on the surface. Without the molecule, the resistance of a nanogap device did not change. However, with the molecule, the conductance of a nanogap device changed reproducibly. Using clean solvent again allowed to repeat the deposition by washing off the molecules, returning the device to its pristine state, and then repeating the deposition. *I-V* curves recorded with the molecules on the device showed an ohmic behavior, with an average resistance of $(90 \pm 85)\,\mathrm{G}\,\Omega$. This high resistance is in contrast to break-junction measurements of OPEs of a similar length, which show a resistance of $\approx 200\,\mathrm{M}\Omega$ for a single molecule [25]. Further studies will have to investigate the observed discrepancies between these results, which can be ascribed, e.g., to diffusive transport along a layer of molecules, diminished conductance of molecules due to imperfect attachment to the CNT electrodes or due to the conformational freedom of molecules in the junction.

In the future, research could aim for atomic-resolution scanning probe studies of the CNT-molecule junctions, to characterize the conductance-limiting factors. However, because the devices are assembled on an insulating silicon oxide surface, an STM characterization of the finished nanogap electrode devices is not possible. Conventional table-top atomic force microscopy (AFM) was shown not to have the required lateral resolution to gain insights into the atomic structure of the nanogaps or their surroundings. Atomic-resolution AFMs exist, however, similar technological challenges compared to atomic-resolution STMs have to be overcome. While the devices of this work are clean on scales assessable by helium-ion microscopy, ultra-clean samples, fabricated in-situ in a UHV system, are preferred. It is likely that to achieve atomic resolution of the carbon nanotube, the substrate beneath it needs to be covered with an atomically flat insulating surface, e.g., exfoliated boron nitride. Furthermore, the limited piezo scan range of atomic-resolution AFMs (as well as STMs) make it impossible to locate micrometer-sized structures on a large substrate without either a complex system of nanoscale markers or an optical or electron microscopic access in the AFM/STM chamber. Still, all these challenges can be overcome.

Another approach to analyzing the defect density of helium-ion-sputtered nanogaps could employ a transmission electron microscope. For this technique the nanogap devices would need to be prepared on electron-transmissive substrates. Subsequently, a comparison between the defects in nanogap electrodes fabricated by current-induced breakdown and helium-ion sputtering would be feasible. The effects of a temperature treatment to eventually heal the defects surrounding a nanogap could be systematically investigated.

The coupling of molecules to the nanogap electrodes is conceivably of prime importance for the conductance of a junction. Variation of the anchor group in molecular synthesis and subsequent integration into nanogap electrodes can provide hints as to the coupling efficiency of different anchor groups to the carbon nanotube sidewall, in the case of π-orbital-mediated coupling. Furthermore, the anchor groups could be designed for covalent bonding to the nanotube. In this case, knowledge of the chemical configuration of the ends of nanogap electrodes after fabrication, i.e., whether they form closed nanotube caps or have dangling carboxylic acids or something entirely different, would be a prerequisite. More complex electrode geometries to contact three or four-terminal molecules could be imagined as well, e.g., by introducing a nanogap at the crossing point of two perpendicularly-aligned carbon nanotubes.

If the coupling of molecules to the nanogaps can be improved, helium-ion sputtering provides a reliable way to fabricate graphitic nanoscale electrodes with an unprecedented size and lowest variability, and will allow the study of other organic and inorganic systems at the single-molecule or few-atom level in the future. Likewise, the STM characterization of OPEs presented in this work can serve as a recipe for other molecules. The integration of functional units such as chromophores into OPE molecules has already been demonstrated, but their electronic properties are still largely unknown. The STM methods of this work could be employed to ascertain, e.g., the energy alignment and structure of molecular orbitals to eventually improve, e.g., their electroluminescence efficiency by modifying the chemical structure through this knowledge. In the case of dye molecules, the extension of the UHV STM with an optical feed-through and integrated optics could allow detailed insights into single-molecule electroluminescence and allow comparisons to the situation of a molecule contacted by a nanogap. We hope that the methods and results presented in this work help pave the way towards combining both of these approaches in future research, particularly concerning molecular wires made from OPE molecules.

Bibliography

[1] C. Thiele, H. Vieker, A. Beyer, B. S. Flavel, F. Hennrich, D. Muñoz Torres, T. R. Eaton, M. Mayor, M. M. Kappes, A. Gölzhäuser, H. v. Löhneysen, and R. Krupke. “Fabrication of carbon nanotube nanogap electrodes by helium ion sputtering for molecular contacts”. *Appl. Phys. Lett.* 104 (2014), 103102. DOI: 10.1063/1.4868097.

[2] C. Thiele, A. Felten, T. J. Echtermeyer, A. C. Ferrari, C. Casiraghi, H. v. Löhneysen, and R. Krupke. “Electron-beam-induced direct etching of graphene”. *Carbon* 64 (2013), 84–91. DOI: 10.1016/j.carbon.2013.07.038.

[3] C. Thiele, M. Engel, F. Hennrich, M. M. Kappes, K.-P. Johnsen, C. G. Frase, H. v. Löhneysen, and R. Krupke. “Controlled fabrication of single-walled carbon nanotube electrodes by electron-beam-induced oxidation”. *Appl. Phys. Lett.* 99 (2011), 173105. DOI: 10.1063/1.3656736.

[4] D. Edelstein, J. Heidenreich, R. Goldblatt, W. Cote, C. Uzoh, N. Lustig, P. Roper, T. McDevitt, W. Motsiff, A. Simon, J. Dukovic, R. Wachnik, H. Rathore, R. Schulz, L. Su, S. Luce, and J. Slattery. “Full copper wiring in a sub-0.25 µm CMOS ULSI technology”. In: *International Electron Devices Meeting, 1997. IEDM '97. Technical Digest.* 1997, 773–776. DOI: 10.1109/IEDM.1997.650496.

[5] D. K. James and J. M. Tour. “Molecular Wires”. *Top. Curr. Chem.* 257 (2005), 33–62. DOI: 10.1007/b136066.

[6] B. Mann and H. Kuhn. “Tunneling through Fatty Acid Salt Monolayers”. *J. Appl. Phys.* 42 (1971), 4398–4405. DOI: 10.1063/1.1659785.

[7] C. J. Muller, J. M. van Ruitenbeek, and L. J. de Jongh. “Experimental observation of the transition from weak link to tunnel junction”. *Physica C: Superconductivity* 191 (1992), 485–504. DOI: 10.1016/0921-4534(92)90947-B.

[8] M. A Reed, C. Zhou, C. J. Muller, T. P. Burgin, and J. M. Tour. “Conductance of a Molecular Junction”. *Science* 278 (1997), 252—254. DOI: 10.1126/science.278.5336.252.

[9] G. Binnig and H. Rohrer. “Scanning apparatus for surface analysis using vacuum-tunnel effect at cryogenic temperatures”. Pat. 1979.

[10] G. Binnig, H. Rohrer, C. Gerber, and E. Weibel. “Surface Studies by Scanning Tunneling Microscopy”. *Phys. Rev. Lett.* 49 (1982), 57–61. DOI: 10.1103/PhysRevLett.49.57.

[11] K. Moshammer, F. Hennrich, and M. M. Kappes. "Selective suspension in aqueous sodium dodecyl sulfate according to electronic structure type allows simple separation of metallic from semiconducting single-walled carbon nanotubes". *Nano Res* 2 (2009), 599—606. DOI: 10.1007/s12274-009-9057-0.

[12] P. Qi, A. Javey, M. Rolandi, Q. Wang, E. Yenilmez, and H. Dai. "Miniature Organic Transistors with Carbon Nanotubes as Quasi-One-Dimensional Electrodes". *J. Am. Chem. Soc.* 126 (2004), 11774–11775. DOI: 10.1021/ja045900k.

[13] X. Guo, J. P. Small, J. E. Klare, Y. Wang, M. S. Purewal, I. W. Tam, B. H. Hong, R. Caldwell, L. Huang, S. O'Brien, J. Yan, R. Breslow, S. J. Wind, J. Hone, P. Kim, and C. Nuckolls. "Covalently Bridging Gaps in Single-Walled Carbon Nanotubes with Conducting Molecules". *Science* 311 (2006), 356–359. DOI: 10.1126/science.1120986.

[14] C. W. Marquardt, S. Grunder, A. Błaszczyk, S. Dehm, F. Hennrich, H. v. Löhneysen, M. Mayor, and R. Krupke. "Electroluminescence from a single nanotube–molecule–nanotube junction". *Nat. Nanotechnol.* 5 (2010), 863—867. DOI: 10.1038/nnano.2010.230.

[15] F. Xiong, A. D. Liao, D. Estrada, and E. Pop. "Low-power switching of phase-change materials with carbon nanotube electrodes". *Science* 332 (2011), 568––570. DOI: 10.1126/science.1201938.

[16] M. Moroni, J. Le Moigne, and S. Luzzati. "Rigid Rod Conjugated Polymers for Nonlinear Optics. 1. Characterization and Linear Optical Properties of Poly(aryleneethynylene) Derivatives". *Macromolecules* 27 (1994), 562–571. DOI: 10.1021/ma00080a034.

[17] J. M. Tour, L. Jones II, D. L. Pearson, J. S. Lamba, T. P. Burgin, G. M. Whitesides, D. L. Allara, A. N. Parikh, and S. V. Atre. "Self- Assembled Monolayers and Multilayers of Conjugated Understanding Attachments between Potential Molecular Wires and Gold Surfaces". *J. Am. Chem. Soc.* 117 (1995), 9529–9534. DOI: 10.1021/ja00142a021.

[18] J. Chen, M. A. Reed, A. M. Rawlett, and J. M. Tour. "Large On-Off Ratios and Negative Differential Resistance in a Molecular Electronic Device". *Science* 286 (1999), 1550–1552. DOI: 10.1126/science.286.5444.1550.

[19] J. Chen, W. Wang, M. a. Reed, a. M. Rawlett, D. W. Price, and J. M. Tour. "Room-temperature negative differential resistance in nanoscale molecular junctions". *Appl. Phys. Lett.* 77 (2000), 1224–1226. DOI: 10.1063/1.1289650.

[20] J. M. Tour, A. M. Rawlett, M. Kozaki, Y. Yao, R. C. Jagessar, S. M. Dirk, D. W. Price, M. A. Reed, C.-w. Zhou, J. Chen, W. Wang, and I. Campbell. "Synthesis and Preliminary Testing of Molecular Wires and Devices". *Chem. Eur. J.* (2001), 5118–5134. DOI: 10.1002/1521-3765(20011203)7:23<5118::AID-CHEM5118>3.0.CO;2-1.

[21] L. A. Bumm, J. J. Arnold, M. T. Cygan, T. D. Dunbar, T. P. Burgin, L. Jones II, D. L. Allara, J. M. Tour, and P. S. Weiss. "Are Single Molecular Wires Conducting?" *Science* 271 (1996), 1705–1707. DOI: 10.1126/science.271.5256.1705.

[22] M. T. Cygan, T. D. Dunbar, J. J. Arnold, L. A. Bumm, N. F. Shedlock, T. P. Burgin, L. Jones II, D. L. Allara, J. M. Tour, and P. S. Weiss. "Insertion, Conductivity, and Structures of Conjugated Organic Oligomers in Self-Assembled Alkanethiol Monolayers on Au{111}". *J. Am. Chem. Soc.* 120 (1998), 2721–2732. DOI: 10.1021/ja973448h.

[23] L. Venkataraman, J. E. Klare, C. Nuckolls, M. S. Hybertsen, and M. L. Steigerwald. "Dependence of single-molecule junction conductance on molecular conformation". *Nature* 442 (2006), 904–907. DOI: 10.1038/nature05037.

[24] X. Xiao, L. A. Nagahara, A. M. Rawlett, and N. Tao. "Electrochemical gate-controlled conductance of single oligo(phenylene ethynylene)s." *J. Am. Chem. Soc.* 127 (2005), 9235–9240. DOI: 10.1021/ja050381m.

[25] Q. Lu, K. Liu, H. Zhang, Z. Du, X. Wang, and F. Wang. "From Tunneling to Hopping: A Comprehensive Investigation of Charge Transport Mechanism in Molecular Junctions Based on Oligo(p-phenylene ethynylene)s". *ACS Nano* 3 (2009), 3861–3868. DOI: 10.1021/nn9012687.

[26] L. Jones II, J. S. Schumm, and J. M. Tour. "Rapid Solution and Solid Phase Syntheses of Oligo(1,4-phenylene ethynylene)s with Thioester Termini: Molecular Scale Wires with Alligator Clips. Derivation of Iterative Reaction Efficiencies on a Polymer Support". *J. Org. Chem.* 62 (1997), 1388–1410. DOI: 10.1021/jo962336q.

[27] A.-a. Dhirani, R. W. Zehner, R. P. Hsung, P. Guyot-Sionnest, and L. R. Sita. "Self-Assembly of Conjugated Molecular Rods: A High-Resolution STM Study". *J. Am. Chem. Soc.* 118 (1996), 3319–3320. DOI: 10.1021/ja953782i.

[28] J.-r. Gong, J.-l. Zhao, S.-b. Lei, L.-j. Wan, Z.-s. Bo, X.-l. Fan, and C.-l. Bai. "Molecular Organization of Alkoxy-Substituted Oligo (phenylene-ethynylene) s Studied by Scanning Tunneling Microscopy". *Langmuir* 19 (2003), 10128–10131. DOI: 10.1021/la034953x.

[29] Z. Mu, X. Yang, Z. Wang, X. Zhang, J. Zhao, and Z. Bo. "Influence of Substituents on Two-Dimensional Ordering of Oligo(phenylene-ethynylene)s - A Scanning Tunneling Microscopy Study". *Langmuir* 20 (2004), 8892–8896. DOI: 10.1021/la048538w.

[30] S.-U. Kim, B.-S. Kim, J.-C. Park, H.-K. Shin, and Y.-S. Kwon. "Electrical properties of single nitro oligophenylene ethynylene (OPE) molecule by using UHV STM". *Current Applied Physics* 6 (2006), 608–611. DOI: 10.1016/j.cap.2005.04.004.

[31] D. Mössinger, S.-S. Jester, E. Sigmund, U. Müller, and S. Höger. "Defined Oligo(p-phenylene-butadiynylene) Rods". *Macromolecules* 42 (2009), 7974–7978. DOI: 10.1021/ma901155d.

[32] F. Maya, A. K. Flatt, M. P. Stewart, D. E. Shen, and J. M. Tour. "Formation and Analysis of Self-Assembled Monolayers from U-Shaped Oligo(phenylene ethynylene)s as Candidates for Molecular Electronics". *Chem. Mater.* 16 (2004), 2987–2997. DOI: 10.1021/cm049504c.

[33] S.-S. Jester, N. Shabelina, S. M. Le Blanc, and S. Höger. "Oligomers and Cyclooligomers of Rigid Phenylene-Ethynylene-Butadiynylenes: Synthesis and Self-Assembled Monolayers". *Angew. Chem. Int. Ed.* 49 (2010), 6101–6105. DOI: 10.1002/anie.201001625.

[34] S.-S. Jester, A. Idelson, D. Schmitz, F. Eberhagen, and S. Höger. "Shape-persistent linear, kinked, and cyclic oligo(phenylene-ethynylene-butadiynylene)s: self-assembled monolayers". *Langmuir* 27 (2011), 8205–15. DOI: 10.1021/la201413h.

[35] S.-S. Jester, D. Schmitz, F. Eberhagen, and S. Höger. "Self-assembled monolayers of clamped oligo(phenylene-ethynylene-butadiynylene)s". *Chem. Commun.* 47 (2011), 8838–8840. DOI: 10.1039/c1cc12435h.

[36] J. M. Seminario, A. G. Zacarias, and J. M. Tour. "Theoretical Study of a Molecular Resonant Tunneling Diode". *J. Am. Chem. Soc.* 122 (2000), 3015–3020. DOI: 10.1021/ja992936h.

[37] C. Majumder, T. Briere, H. Mizuseki, and Y. Kawazoe. "Molecular Resistance in a Molecular Diode: A Case Study of the Substituted Phenylethynyl Oligomer". *J. Phys. Chem. A* 106 (2002), 7911–7914. DOI: 10.1021/jp0258560.

[38] D. Vonlanthen, A. Mishchenko, M. Elbing, M. Neuburger, T. Wandlowski, and M. Mayor. "Chemically Controlled Conductivity: Torsion-Angle Dependence in a Single-Molecule Biphenyldithiol Junction". *Angew. Chem. Int. Ed.* 48 (2009), 8886–8890. DOI: 10.1002/anie.200903946.

[39] Y. Li, J. Zhao, and G. Yin. "Theoretical investigations of oligo(phenylene ethylene) molecular wire: Effects from substituents and external electric field". *Computational Materials Science* 39 (2007), 775–781. DOI: 10.1016/j.commatsci.2006.09.010.

[40] J.-Q. Lu, J. Wu, H. Chen, W. Duan, B.-L. Gu, and Y. Kawazoe. "Electronic transport mechanism of a molecular electronic device: structural effects and terminal atoms". *Physics Letters A* 323 (2004), 154–158. DOI: 10.1016/j.physleta.2004.01.055.

[41] S. Gotovac, Y. Hattori, D. Noguchi, J.-I. Miyamoto, M. Kanamaru, S. Utsumi, H. Kanoh, and K. Kaneko. "Phenanthrene adsorption from solution on single wall carbon nanotubes". *J. Phys. Chem. B* 110 (2006), 16219–16224. DOI: 10.1021/jp0611830.

[42] S. Gotovac, C.-M. Yang, Y. Hattori, K. Takahashi, H. Kanoh, and K. Kaneko. "Adsorption of polyaromatic hydrocarbons on single wall carbon nanotubes of different functionalities and diameters". *J. Colloid Interface Sci.* 314 (2007), 18–24. DOI: 10.1016/j.jcis.2007.04.080.

[43] S. Grunder, D. Muñoz Torres, C. Marquardt, A. Błaszczyk, R. Krupke, and M. Mayor. "Synthesis and Optical Properties of Molecular Rods Comprising a Central Core-Substituted Naphthalenediimide Chromophore for Carbon Nanotube Junctions". *Eur. J. Org.Chem.* 2011 (2011), 478–496. DOI: 10.1002/ejoc.201001415.

[44] R. Wiesendanger. *Scanning Probe Microscopy and Spectroscopy.* Cambridge University Press, Cambridge, UK, 1994.

[45] R. Hamers and D. Padowitz. In: *Scanning Probe Microscopy and Spectroscopy: Theory, Techniques, and Applications, 2nd Edition.* Ed. by D. Bonnell. Wiley-VCH, New York, USA, 2001. Chap. Methods of Tunneling Spectroscopy with STM.

[46] J. Bardeen. "Tunnelling from a Many-Particle Point of View". *Phys. Rev. Lett.* 6 (1961), 57–59. DOI: 10.1103/PhysRevLett.6.57.

[47] J. Tersoff and D. R. Hamann. "Theory and Application for the Scanning Tunneling Microscope". *Phys. Rev. Lett.* 50 (1983), 1998–2001. DOI: 10.1103/PhysRevLett.50.1998.

[48] J. Tersoff and D. R. Hamann. "Theory of the scanning tunneling microscope". *Phys. Rev. B* 31 (1985), 805–813. DOI: 10.1103/PhysRevB.31.805.

[49] J. Stroscio, R. Feenstra, and A. Fein. "Electronic structure of the Si (111) 2 $\times$ 1 surface by scanning-tunneling microscopy". *Phys. Rev. Lett.* 57 (1986), 2579–2582. DOI: 10.1103/PhysRevLett.57.2579.

[50] T. Trappmann. "Rastertunnelmikroskopie und -spektroskopie an phosphordotiertem Silizium und an dünnen Gadolinium-Schichten". PhD thesis. Universität Karlsruhe (TH), 1996.

[51] J. Repp, G. Meyer, S. Stojković, A. Gourdon, and C. Joachim. "Molecules on Insulating Films: Scanning-Tunneling Microscopy Imaging of Individual Molecular Orbitals". *Phys. Rev. Lett.* 94 (2005), 026803. DOI: 10.1103/PhysRevLett.94.026803.

[52] C. J. Villagomez, T. Zambelli, S. Gauthier, A. Gourdon, S. Stojkovic, and C. Joachim. "STM images of a large organic molecule adsorbed on a bare metal substrate or on a thin insulating layer: Visualization of HOMO and LUMO". *Surface Sci.* 603 (2009), 1526–1532. DOI: 10.1016/j.susc.2008.10.057.

[53] P. Liljeroth, J. Repp, and G. Meyer. "Current-induced hydrogen tautomerization and conductance switching of naphthalocyanine molecules". *Science* 317 (2007), 1203–1206. DOI: 10.1126/science.1144366.

[54] L. Gross, N. Moll, F. Mohn, A. Curioni, G. Meyer, F. Hanke, and M. Persson. "High-Resolution Molecular Orbital Imaging Using a p-Wave STM Tip". *Phys. Rev. Lett.* 107 (2011), 086101. DOI: 10.1103/PhysRevLett.107.086101.

[55] L. Gross, F. Mohn, N. Moll, P. Liljeroth, and G. Meyer. "The chemical structure of a molecule resolved by atomic force microscopy". *Science* 325 (2009), 1110–1114. DOI: 10.1126/science.1176210.

[56] D. G. de Oteyza, P. Gorman, Y.-C. Chen, S. Wickenburg, A. Riss, D. J. Mowbray, G. Etkin, Z. Pedramrazi, H.-Z. Tsai, A. Rubio, M. F. Crommie, and F. R. Fischer. "Direct imaging of covalent bond structure in single-molecule chemical reactions". *Science* 340 (2013), 1434–1437. DOI: 10.1126/science.1238187.

[57] R. Temirov, S. Soubatch, O. Neucheva, a. C. Lassise, and F. S. Tautz. "A novel method achieving ultra-high geometrical resolution in scanning tunnelling microscopy". *New J. Phys.* 10 (2008), 053012. DOI: 10.1088/1367-2630/10/5/053012.

[58] C. Weiss, C. Wagner, C. Kleimann, M. Rohlfing, F. S. Tautz, and R. Temirov. "Imaging Pauli Repulsion in Scanning Tunneling Microscopy". *Phys. Rev. Lett.* 105 (2010), 086103. DOI: 10.1103/PhysRevLett.105.086103.

[59] C. Weiss, C. Wagner, R. Temirov, and F. S. Tautz. "Direct Imaging of Intermolecular Bonds in Scanning Tunneling Microscopy". *J. Am. Chem. Soc.* 132 (2010), 11864–11865. DOI: 10.1021/ja104332t.

[60] J. I. Martínez, E. Abad, C. González, F. Flores, and J. Ortega. "Improvement of Scanning Tunneling Microscopy Resolution with H-Sensitized Tips". *Phys. Rev. Lett.* 108 (2012), 246102. DOI: 10.1103/PhysRevLett.108.246102.

[61] C. Wöll, S. Chiang, R. J. Wilson, and P. H. Lippel. "Determination of atom positions at stacking-fault dislocations on Au(111) by scanning tunneling microscopy". *Phys. Rev. B* 39 (1989), 7988–7991. DOI: 10.1103/PhysRevB.39.7988.

[62] J. V. Barth, H. Brune, G. Ertl, and R. J. Behm. "Scanning tunneling microscopy observations on the reconstructed Au(111) surface: Atomic structure, long-range superstructure, rotational domains, and surface defects". *Phys. Rev. B* 42 (1990), 9307–9318. DOI: 10.1103/PhysRevB.42.9307.

[63] Y. Wang, N. Hush, and J. Reimers. "Simulation of the Au(111)-($22 \times \sqrt{3}$) surface reconstruction". *Phys. Rev. B* 75 (2007), 233416. DOI: 10.1103/PhysRevB.75.233416.

[64] R. G. Musket, W. McLean, A. Colmenares, D. M. Makowiecki, and W. J. Siekhaus. "Preparation of Atomically Clean Surfaces of Selected Elements - A Review". *Applications of Surface Science* 10 (1982), 143–207. DOI: 10.1016/0378-5963(82)90142-8.

[65] P. Heimann and H. Neddermeyer. "Ultraviolet photoemission from single crystals and the bandstructure of gold". *J. Phys. F: Met. Phys.* 7 (1977), L37–L42. DOI: 10.1088/0305-4608/7/1/008.

[66] S. Narasimhan and D. Vanderbilt. "Elastic Stress Domains and the Herringbone Reconstruction on Au(111)". *Phys. Rev. Lett.* 69 (1992), 1564–1568. DOI: 10.1103/PhysRevLett.69.1564.

[67] W. Chen, V. Madhavan, T. Jamneala, and M. F. Crommie. "Scanning Tunneling Microscopy Observation of an Electronic Superlattice at the Surface of Clean Gold". *Phys. Rev. Lett.* 80 (1998), 1469–1472. DOI: 10.1103/PhysRevLett.80.1469.

[68] L. Bürgi, H. Brune, and K. Kern. "Imaging of Electron Potential Landscapes on Au(111)". *Phys. Rev. Lett.* 89 (2002), 176801. DOI: 10.1103/PhysRevLett.89.176801.

[69] N. Takeuchi, C. T. Chan, and K. M. Ho. "Au (111): A theoretical study of the surface reconstruction and the surface electronic structure". *Phys. Rev. B* 43 (1991), 13899–13906. DOI: 10.1103/PhysRevB.43.13899.

[70] W. Kohn. "Nobel Lecture: Electronic structure of matter—wave functions and density functionals". *Rev. Mod. Phys.* 71 (1999), 1253–1266. DOI: 10.1103/RevModPhys.71.1253.

[71] K. Capelle. *A bird's-eye view of density-functional theory.* 2002. arXiv: cond-mat/0211443.

[72] P. Hohenberg and W. Kohn. "Inhomogeneous Electron Gas". *Phys. Rev. B* 136 (1964), 864–871. DOI: 10.1103/PhysRev.136.B864.

[73] W. Kohn and L. J. Sham. "Self-Consistent Equations Including Exchange and Correlation Effects". *Phys. Rev. A* 140 (1965), 1133–1138. DOI: 10.1103/PhysRev.140.A1133.

[74] A. D. Becke. "Density-functional exchange-energy approximation with correct asymptotic behavior". *Phys. Rev. A* 38 (1988), 3098–3100. DOI: 10.1103/PhysRevA.38.3098.

[75] J. P. Perdew. "Density-functional approximation for the correlation energy of the inhomogenenous electron gas". *Phys. Rev. B* 33 (1986), 8822–8824. DOI: 10.1103/PhysRevB.33.8822.

[76] A. D. Becke. "Density-functional thermochemistry. III. The role of exact exchange". *J. Chem. Phys.* 98 (1993), 5648. DOI: 10.1063/1.464913.

[77] C. Lee, W. Yang, and R. G. Parr. "Development of the Colle-Salvetti correlation-energy formula into a functional of the electron density". *Phys. Rev. B* 37 (1988), 785–789. DOI: 10.1103/PhysRevB.37.785.

[78] S. H. Vosko, L. Wilk, and M. Nusair. "Accurate spin-dependent electron liquid correlation energies for local spin density calculations: a critical analysis". *Can. J. Phys.* 58 (1980), 1200–1211. DOI: 10.1139/p80-159.

[79] W. J. Hehre, R. Ditchfield, and J. A. Pople. “Self—Consistent Molecular Orbital Methods. XII. Further Extensions of Gaussian—Type Basis Sets for Use in Molecular Orbital Studies of Organic Molecules”. *J. Chem. Phys.* 56 (1972), 2257. DOI: 10.1063/1.1677527.

[80] R. Ditchfield, W. J. Hehre, and J. A. Pople. “Self-Consistent Molecular-Orbital Methods. IX. An Extended Gaussian-Type Basis for Molecular-Orbital Studies of Organic Molecules”. *J. Chem. Phys.* 54 (1971), 724–728. DOI: 10.1063/1.1674902.

[81] A. Schäfer, H. Horn, and R. Ahlrichs. “Fully optimized contracted Gaussian basis sets for atoms Li to Kr”. *J. Chem. Phys.* 97 (1992), 2571–2577. DOI: 10.1063/1.463096.

[82] K. Eichkorn, F. Weigend, O. Treutler, and R. Ahlrichs. “Auxiliary basis sets for main row atoms and transition metals and their use to approximate Coulomb potentials”. *Theor. Chim. Acta.* 97 (1997), 119–124. DOI: 10.1007/s002140050244.

[83] S. Iijima. “Helical microtubules of graphitic carbon”. *Nature* 354 (1991), 56––58. DOI: 10.1038/354056a0.

[84] P. Wallace. “The Band Theory of Graphite”. *Phys. Rev.* 71 (1947), 622—634. DOI: 10.1103/PhysRev.71.622.

[85] H. P. Boehm, A Clauss, G. Fischer, and U Hofmann. “Dünnste Kohlenstoff-Folien”. *Z. Naturforsch.* 17 b (1962), 150–153.

[86] K. S. Novoselov, A. K. Geim, S. V. Morozov, D. Jiang, Y. Zhang, S. V. Dubonos, I. V. Grigorieva, and A. A. Firsov. “Electric Field Effect in Atomically Thin Carbon Films”. *Science* 306 (2004), 666–669. DOI: 10.1126/science.1102896.

[87] J.-C. Charlier, X. Blase, and S. Roche. “Electronic and transport properties of nanotubes”. *Rev. Mod. Phys.* 79 (2007), 677–732. DOI: 10.1103/RevModPhys.79.677.

[88] F. Bonaccorso, Z. Sun, T. Hasan, and A. C. Ferrari. “Graphene photonics and optoelectronics”. *Nat. Photon.* 4 (2010), 611–622. DOI: 10.1038/nphoton.2010.186.

[89] A. H. Castro Neto, F. Guinea, N. M. R. Peres, K. S. Novoselov, and A. K. Geim. “The electronic properties of graphene”. *Rev. Mod. Phys.* 81 (2009), 109–162. DOI: 10.1103/RevModPhys.81.109.

[90] J. Kong, A. M. Cassell, and H. Dai. “Chemical vapor deposition of methane for single-walled carbon nanotubes”. *Chem. Phys. Lett.* 292 (1998), 567–574. DOI: 10.1016/S0009-2614(98)00745-3.

[91] C. Journet, W. K. Maser, P. Bernier, A. Loiseau, M. L. la Chapelle, S. Lefrant, P. Deniard, R. Leek, and J. E. Fischer. "Large-scale production of single-walled carbon nanotubes by the electric-arc technique". *Nature* 388 (1997), 756–758. DOI: 10.1038/41972.

[92] S. Lebedkin, P. Schweiss, B. Renker, S. Malik, F. Hennrich, M. Neumaier, C. Stoermer, and M. M. Kappes. "Single-wall carbon nanotubes with diameters approaching 6 nm obtained by laser vaporization". *Carbon* 40 (2002), 417—423. DOI: 10.1016/S0008-6223(01)00119-1.

[93] J. Prasek, J. Drbohlavova, J. Chomoucka, J. Hubalek, O. Jasek, V. Adam, and R. Kizek. "Methods for carbon nanotubes synthesis—review". *J. Mater. Chem.* 21 (2011), 15872. DOI: 10.1039/c1jm12254a.

[94] B. S. Flavel, K. E. Moore, M. Pfohl, M. M. Kappes, and F. Hennrich. "Separation of single-walled carbon nanotubes with a gel permeation chromatography system". *ACS Nano* 8 (2014), 1817–1826. DOI: 10.1021/nn4062116.

[95] H. A. Pohl. "The Motion and Precipitation of Suspensoids in Divergent Electric Fields". *J. Appl. Phys.* 22 (1951), 869–871. DOI: 10.1063/1.1700065.

[96] C. W. Marquardt, S. Blatt, F. Hennrich, H. v. Löhneysen, and R. Krupke. "Probing dielectrophoretic force fields with metallic carbon nanotubes". *Appl. Phys. Lett.* 89 (2006), 183117. DOI: 10.1063/1.2372771.

[97] K. Yamamoto, S. Akita, and Y. Nakayama. "Orientation and purification of carbon nanotubes using ac electrophoresis". *J. Phys. D: Appl. Phys.* 31 (1998), L34–L36. DOI: 10.1088/0022-3727/31/8/002.

[98] A. Vijayaraghavan, S. Blatt, D. Weissenberger, M. Oron-Carl, F. Hennrich, D. Gerthsen, H. Hahn, and R. Krupke. "Ultra-Large-Scale Directed Assembly of Single-Walled Carbon Nanotube Devices". *Nano Lett.* 7 (2007), 1556–1560. DOI: 10.1021/nl0703727.

[99] S. Shekhar, P. Stokes, and S. Khondaker. "Ultrahigh Density Alignment of Carbon Nanotube Arrays by Dielectrophoresis". *ACS Nano* 5 (2011), 1739–1746. DOI: 10.1021/nn102305z.

[100] R. Krupke, F. Hennrich, M. M. Kappes, and H. v. Löhneysen. "Surface Conductance Induced Dielectrophoresis of Semiconducting Single-Walled Carbon Nanotubes". *Nano Lett.* 4 (2004), 1395–1399. DOI: 10.1021/nl0493794.

[101] R. L. Stewart. "Insulating Films Formed Under Electron and Ion Bombardment". *Phys. Rev.* 45 (1934), 488–490. DOI: 10.1103/PhysRev.45.488.

[102] R. W. Christy. "Formation of Thin Polymer Films by Electron Bombardment". *J. Appl. Phys.* 31 (1960), 1680. DOI: 10.1063/1.1735915.

[103] A. N. Broers, W. W. Molzen, J. J. Cuomo, and N. D. Wittels. "Electron-beam fabrication of 80-Å metal structures". *Appl. Phys. Lett.* 29 (1976), 596. DOI: 10.1063/1.89155.

[104] J. Linders, H. Niedrig, N. Ram, and H. Koch. "Production of lead microcones by contamination lithography". *Radiation Effects* 88 (1986), 105–112. DOI: 10.1080/00337578608207501.

[105] M. Yamamoto, M. Sato, H. Kyogoku, K. Aita, Y. Nakagawa, A. Yasaka, R. Takasawa, and O. Hattori. *Submicron Mask Repair Using Focused Ion Beam Technology.* 1986. DOI: 10.1117/12.963674.

[106] S. Matsui and K. Mori. "New selective deposition technology by electron beam induced surface reaction". *J. Vac. Sci. Technol. B* 4 (1986), 299–304. DOI: 10.1116/1.583317.

[107] H. P. Koops, R. Weiel, D. P. Kern, and T. H. Baum. "High-resolution electron-beam induced deposition". *J. Vac. Sci. Technol. B* 6 (1988), 477–481. DOI: 10.1116/1.584045.

[108] R. J. Young, J. R. A. Cleaver, and H. Ahmed. "Characteristics of gas-assisted focused ion beam etching". *J. Vac. Sci. Technol. B* 11 (1993), 234–241. DOI: 10.1116/1.586708.

[109] I. Utke, P. Hoffmann, and J. Melngailis. "Gas-assisted focused electron beam and ion beam processing and fabrication". *J. Vac. Sci. Technol. B* 26 (2008), 1197–1276. DOI: 10.1116/1.2955728.

[110] M. Knoll and E. Ruska. "Das Elektronenmikroskop". *Z. Physik* 78 (1932), 318–339. DOI: 10.1007/BF01342199.

[111] M. von Ardenne. *Elektronen-Übermikroskopie.* Springer Verlag, Berlin, 1940. DOI: 10.1007/978-3-642-47348-7.

[112] H. Seiler. "Secondary electron emission in the scanning electron microscope". *J. Appl. Phys.* 54 (1983), R1–R18. DOI: 10.1063/1.332840.

[113] E. Schreiber and H.-J. Fitting. "Monte Carlo simulation of secondary electron emission from the insulator SiO_2". *Journal of Electron Spectroscopy and Related Phenomena* 124 (2002), 25–37. DOI: 10.1016/S0368-2048(01)00368-1.

[114] T. D. Yuzvinsky, A. M. Fennimore, W. Mickelson, C. Esquivias, and A. Zettl. "Precision cutting of nanotubes with a low-energy electron beam". *Appl. Phys. Lett.* 86 (2005), 053109. DOI: 10.1063/1.1857081.

[115] P. Liu, F. Arai, and T. Fukuda. "Cutting of carbon nanotubes assisted with oxygen gas inside a scanning electron microscope". *Appl. Phys. Lett.* 89 (2006), 113104. DOI: 10.1063/1.2348779.

[116] P. S. Spinney, D. G. Howitt, S. D. Collins, and R. L. Smith. "Electron beam stimulated oxidation of carbon". *Nanotechnology* 20 (2009), 465301. DOI: 10.1088/0957-4484/20/46/465301.

[117] B. W. Smith and D. E. Luzzi. "Electron irradiation effects in single wall carbon nanotubes". *J. Appl. Phys.* 90 (2001), 3509–3515. DOI: 10.1063/1.1383020.

[118] D. Rapp and P. Englander-Golden. "Total Cross Sections for Ionization and Attachment in Gases by Electron Impact. I. Positive Ionization". *J. Chem. Phys.* 43 (1965), 1464–1479. DOI: 10.1063/1.1696957.

[119] B. L. Schram, F. J. De Heer, M. H. Van Der Wiel, and J. Kistemaker. "Ionization Cross Sections For Electrons (0.6-20keV) in Noble and Diatomic Gases". *Physica* 31 (1965), 94–112. DOI: 10.1016/0031-8914(65)90109-6.

[120] E. C. Zipf. "The ionization of atomic oxygen by electron impact". *Planet. Space. Sci.* 33 (1985), 1303–1307. DOI: 10.1016/0032-0633(85)90008-X.

[121] W. R. Thompson, M. B. Shah, and H. B. Gilbody. "Single and double ionization of atomic oxygen by electron impact". *J. Phys. B: At. Mol. Opt. Phys* 28 (1995), 1321–1330. DOI: 10.1088/0953-4075/28/7/023.

[122] Y.-K. Kim and J.-P. Desclaux. "Ionization of carbon, nitrogen, and oxygen by electron impact". *Phys. Rev. A* 66 (2002), 012708. DOI: 10.1103/PhysRevA.66.012708.

[123] D. Rapp and D. D. Briglia. "Total Cross Sections for Ionization and Attachment in Gases by Electron Impact. II. Negative-Ion Formation". *J. Chem. Phys.* 43 (1965), 1480–1489. DOI: 10.1063/1.1696958.

[124] R. Hill, J. A. Notte, and L. Scipioni. "Chapter 2 - Scanning Helium Ion Microscopy". In: *Advances in Imaging and Electron Physics.* Ed. by P. W. Hawkes. Vol. 170. Elsevier, 2012, 65–148. DOI: 10.1016/B978-0-12-394396-5.00002-6.

[125] K. Ohya, T. Yamanaka, K. Inai, and T. Ishitani. "Comparison of secondary electron emission in helium ion microscope with gallium ion and electron microscopes". *Nucl. Instrum. Methods Phys. Res. B* 267 (2009), 584–589. DOI: 10.1016/j.nimb.2008.11.003.

[126] R. Ramachandra, B. Griffin, and D. Joy. "A model of secondary electron imaging in the helium ion scanning microscope". *Ultramicroscopy* 109 (2009), 748–757. DOI: 10.1016/j.ultramic.2009.01.013.

[127] D. C. Bell. "Contrast Mechanisms and Image Formation in Helium Ion Microscopy". *Microsc. Microanal.* 15 (2009), 147–153. DOI: 10.1017/S1431927609090138.

[128] J. Morgan, J. Notte, R. Hill, and B. Ward. "An Introduction to the Helium Ion Microscope". *Microscopy Today* 14 (2006).

[129] E. W. Müller. "Das Feldionenmikroskop". *Z. Physik* 131 (1951), 136–142. DOI: 10.1007/BF01329651.

[130] E. W. Müller and K. Bahadur. "Field Ionization of Gases at a Metal Surface and the Resolution of the Field Ion Microscope". *Phys. Rev.* 102 (1956), 624. DOI: 10.1103/PhysRev.102.624.

[131] D. Winston, B. M. Cord, B. Ming, D. C. Bell, W. F. DiNatale, L. a. Stern, a. E. Vladar, M. T. Postek, M. K. Mondol, J. K. W. Yang, and K. K. Berggren. "Scanning-helium-ion-beam lithography with hydrogen silsesquioxane resist". *J. Vac. Sci. Technol. B* 27 (2009), 2702. DOI: 10.1116/1.3250204.

[132] D. C. Bell, M. C. Lemme, L. a. Stern, J. R. Williams, and C. M. Marcus. "Precision cutting and patterning of graphene with helium ions". *Nanotechnology* 20 (2009), 455301. DOI: 10.1088/0957-4484/20/45/455301.

[133] M. C. Lemme, D. C. Bell, J. R. Williams, L. a. Stern, B. W. H. Baugher, P. Jarillo-Herrero, and C. M. Marcus. "Etching of graphene devices with a helium ion beam". *ACS Nano* 3 (2009), 2674–2676. DOI: 10.1021/nn900744z.

[134] D. Fox, Y. B. Zhou, A. O'Neill, S. Kumar, J. J. Wang, J. N. Coleman, G. S. Duesberg, J. F. Donegan, and H. Z. Zhang. "Helium ion microscopy of graphene: beam damage, image quality and edge contrast". *Nanotechnology* 24 (2013), 335702. DOI: 10.1088/0957-4484/24/33/335702.

[135] M. M. Marshall, J. Yang, and A. R. Hall. "Direct and transmission milling of suspended silicon nitride membranes with a focused helium ion beam". *Scanning* 34 (2012), 101–106. DOI: 10.1002/sca.21003.

[136] O. Scholder, K. Jefimovs, I. Shorubalko, C. Hafner, U. Sennhauser, and G.-L. Bona. "Helium focused ion beam fabricated plasmonic antennas with sub-5 nm gaps". *Nanotechnology* 24 (2013), 395301. DOI: 10.1088/0957-4484/24/39/395301.

[137] Y. Wang, M. Abb, S. a. Boden, J. Aizpurua, C. H. de Groot, and O. L. Muskens. "Ultrafast Nonlinear Control of Progressively Loaded, Single Plasmonic Nanoantennas Fabricated Using Helium Ion Milling". *Nano Lett.* 13 (2013), 5647–5653. DOI: 10.1021/nl403316z.

[138] D. Fox, Y. Chen, C. C. Faulkner, and H. Zhang. "Nano-structuring, surface and bulk modification with a focused helium ion beam". *Beilstein J. Nanotechnol.* 3 (2012), 579–585. DOI: 10.3762/bjnano.3.67.

[139] O. Lehtinen, J. Kotakoski, A. V. Krasheninnikov, and J. Keinonen. "Cutting and controlled modification of graphene with ion beams". *Nanotechnology* 22 (2011), 175306. DOI: 10.1088/0957-4484/22/17/175306.

[140] K. A. Dössel. "Rastertunnelmikroskopie und -spektroskopie an multiterminierten Metallkomplexmolekülen auf Au(111)- und Ag(111)-Oberflächen". PhD thesis. Karlsruher Institut für Technologie, 2011.

[141] M. Rossberg, W. Lendle, G. Pfleiderer, A. Tögel, T. R. Torkelson, and K. K. Beutel. "Chloromethanes". In: *Ullmann's Encyclopedia of Industrial Chemistry*. Vol. 9. Wiley-VCH Verlag GmbH & Co. KGaA, 2011. ISBN: 9783527306732. DOI: 10.1002/14356007.a06_233.pub3.

[142] R. D. Deegan, O. Bakajin, T. F. Dupont, G. Huber, S. R. Nagel, and T. A. Witten. "Capillary flow as the cause of ring stains from dried liquid drops". *Nature* 389 (1997), 827–829. DOI: 10.1038/39827.

[143] T. A. Jung, R. R. Schlittler, J. K. Gimzewski, H. Tang, and C. Joachim. "Controlled Room-Temperature Positioning of Individual Molecules: Molecular Flexure and Motion". *Science* 271 (1996), 181–184. DOI: 10.1126/science.271.5246.181.

[144] W. M. H. Sachtler, G. J. H. Dorgelo, and A. A. Holscher. "The work function of gold". *Surface Sci.* 5 (1966), 221–229. DOI: 10.1016/0039-6028(66)90083-5.

[145] M. Büttiker, Y. Imry, R. Landauer, and S. Pinhas. "Generalized many-channel conductance formula with application to small rings". *Phys. Rev. B* 31 (1985), 6207–6215. DOI: 10.1103/PhysRevB.31.6207.

[146] D. M. Eigler, C. P. Lutz, and W. E. Rudge. "An atomic switch realized with the scanning tunnelling microscope". *Nature* 352 (1991), 600–603. DOI: 10.1038/352600a0.

[147] M. Brandbyge and P. Hedegård. "Theory of the Eigler switch". *Phys. Rev. Lett.* 72 (1994), 2919–2922. DOI: 10.1103/PhysRevLett.72.2919.

[148] S. Gao, M. Persson, and B. I. Lundqvist. "Theory of atom transfer with a scanning tunneling microscope". *Phys. Rev. B* 55 (1997), 4825–4836. DOI: 10.1103/PhysRevB.55.4825.

[149] B. Stipe, M. Rezaei, W. Ho, S. Gao, M. Persson, and B. Lundqvist. "Single-Molecule Dissociation by Tunneling Electrons". *Phys. Rev. Lett.* 78 (1997), 4410–4413. DOI: 10.1103/PhysRevLett.78.4410.

[150] Z. J. Donhauser, B. a. Mantooth, K. F. Kelly, L. a. Bumm, J. D. Monnell, J. J. Stapleton, D. W. Price Jr, a. M. Rawlett, D. L. Allara, J. M. Tour, and P. S. Weiss. "Conductance switching in single molecules through conformational changes". *Science* 292 (2001), 2303–2307. DOI: 10.1126/science.1060294.

[151] F. Elste, G. Weick, C. Timm, and F. Oppen. "Current-induced conformational switching in single-molecule junctions". *Applied Physics A* 93 (2008), 345–354. DOI: 10.1007/s00339-008-4826-2.

[152] S. Tikhodeev and H. Ueba. "Relation between inelastic electron tunneling and vibrational excitation of single adsorbates on metal surfaces". *Phys. Rev. B* 70 (2004), 125414. DOI: 10.1103/PhysRevB.70.125414.

[153] T. Brumme, O. a. Neucheva, C. Toher, R. Gutiérrez, C. Weiss, R. Temirov, A. Greuling, M. Kaczmarski, M. Rohlfing, F. S. Tautz, and G. Cuniberti. "Dynamical bistability of single-molecule junctions: A combined experimental and theoretical study of PTCDA on Ag(111)". *Phys. Rev. B* 84 (2011), 115449. DOI: 10.1103/PhysRevB.84.115449.

[154] Y. E. Shchadilova, S. G. Tikhodeev, M. Paulsson, and H. Ueba. "Rotation of a Single Acetylene Molecule on Cu(001) by Tunneling Electrons in STM". *Phys. Rev. Lett.* 111 (2013), 186102. DOI: 10.1103/PhysRevLett.111.186102.

[155] *TURBOMOLE V6.4 2012, 1989-2007, TURBOMOLE GmbH, since 2007; available from* http://www.turbomole.com.

[156] M. D. Hanwell, D. E. Curtis, D. C. Lonie, T. Vandermeersch, E. Zurek, and G. R. Hutchison. "Avogadro: an advanced semantic chemical editor, visualization, and analysis platform". *J. Cheminform.* 4 (2012), 1–17. DOI: 10.1186/1758-2946-4-17.

[157] T. A. Halgren. "Merck Molecular Force Field. I. Basis, Form, Scope, Parameterization, and Performance of MMFF94". *J. Comp. Chem.* 17 (1996), 490–519. DOI: 10.1002/(SICI)1096-987X(199604)17:5/6<490::AID-JCC1>3.0.CO;2-P.

[158] D. Andrae, U. Häußermann, M. Dolg, H. Stoll, and H. Preuß. "Energy-adjusted ab initio pseudopotentials for the second and third row transition elements". *Theor. Chim. Acta.* 77 (1990), 123–141. DOI: 10.1007/BF01114537.

[159] S. Grimme, J. Antony, S. Ehrlich, and H. Krieg. "A consistent and accurate ab initio parametrization of density functional dispersion correction (DFT-D) for the 94 elements H-Pu." *J. Chem. Phys.* 132 (2010), 154104. DOI: 10.1063/1.3382344.

[160] S. Grimme, S. Ehrlich, and L. Goerigk. "Effect of the Damping Function in Dispersion Corrected Density Functional Theory". *J. Comp. Chem.* 32 (2011), 1456–1465. DOI: 10.1002/jcc.21759.

[161] S. Blatt, F. Hennrich, H. v. Löhneysen, M. M. Kappes, A. Vijayaraghavan, and R. Krupke. "Influence of Structural and Dielectric Anisotropy on the Dielectrophoresis of Single-Walled Carbon Nanotubes". *Nano Letters* 7 (2007), 1960–1966. DOI: 10.1021/nl0706751.

[162] A. Vijayaraghavan, S. Blatt, C. Marquardt, S. Dehm, R. Wahi, F. Hennrich, and R. Krupke. "Imaging electronic structure of carbon nanotubes by voltage-contrast scanning electron microscopy". *Nano Res* 1 (2008), 321–332. DOI: 10.1007/s12274-008-8034-3.

[163] C.-L. Zhang and H.-S. Shen. "Self-healing in defective carbon nanotubes: a molecular dynamics study". *J. Phys.: Condens. Matter* 19 (2007), 386212. DOI: 10.1088/0953-8984/19/38/386212.

[164] O. Suekane, A. Nagataki, and Y. Nakayama. "Current-induced curing of defective carbon nanotubes". *Appl. Phys. Lett.* 89 (2006), 183110. DOI: 10.1063/1.2372749.

[165] S. Shekhar, M. Erementchouk, M. N. Leuenberger, and S. I. Khondaker. "Correlated electrical breakdown in arrays of high density aligned carbon nanotubes". *Appl. Phys. Lett.* 98 (2011), 243121. DOI: 10.1063/1.3600664.

[166] D. Estrada and E. Pop. "Imaging dissipation and hot spots in carbon nanotube network transistors". *Appl. Phys. Lett.* 98 (2011), 073102. DOI: 10.1063/1.3549297.

[167] C. W. Marquardt, S. Dehm, A. Vijayaraghavan, S. Blatt, F. Hennrich, and R. Krupke. "Reversible Metal - Insulator Transitions in Metallic Single-Walled Carbon Nanotubes". *Nano Lett.* 8 (2008), 2767–2772.

[168] K.-P. Johnsen, C. G. Frase, H. Bosse, and D. Gnieser. *SEM image modeling using the modular Monte Carlo model MCSEM.* 2010. DOI: 10.1117/12.846543.

[169] W. Kim, A. Javey, R. Tu, J. Cao, Q. Wang, and H. Dai. "Electrical contacts to carbon nanotubes down to 1nm in diameter". *Appl. Phys. Lett.* 87 (2005), 173101. DOI: 10.1063/1.2108127.

[170] J. G. Simmons. "Generalized Formula for the Electric Tunnel Effect between Similar Electrodes Separated by a Thin Insulating Film". *J. Appl. Phys.* 34 (1963), 1793–1803. DOI: 10.1063/1.1702682.

[171] M. Shiraishi and M. Ata. "Work function of carbon nanotubes". *Carbon* 39 (2001), 1913–1917. DOI: 10.1016/S0008-6223(00)00322-5.

Acknowledgments

I would like to thank many people for their support, both scientifically and morally:

- Prof. Hilbert von Löhneysen for giving me the opportunity to work in this exciting field, for all the helpful discussions and his support in writing scientific papers and this thesis.
- Prof. Wulf Wulfhekel for co-refereeing this work, for helping us in designing our STM, and his help in interpreting the STM results.
- Prof. Ralph Krupke and Dr. Maya Lukas for their supervision, for all their ideas, help and encouragement and all the fruitful discussions we had around the coffee table and in the labs. This work would not have been possible without either of them.
- Dr. Lukas Gerhard, Dr. Lei Zhang and Kevin Edelmann for their help in setting up and running the new STM.
- Prof. Ferdinand Evers, Dr. Karin Fink and Dr. Timo Strunk for introducing me to density functional theory calculations and discussion of results.
- Prof. Marcel Mayor, Dr. Thomas Eaton, Dr. David Muñoz Torres and Dr. Michal Valášek for synthesis of the OPE molecules and providing me with ultra-clean solvent.
- Dr. Frank Hennrich and Dr. Benjamin Scott Flavel for supplying me with sorted carbon nanotubes.
- Dr. Carl Georg Frase and Dr. Klaus-Peter Johnsen for the Monte Carle simulations of secondary-electron generation in silicon substrates.
- Prof. Armin Gölzhäuser, Dr. André Beyer and Henning Vieker for our cooperative efforts of cutting carbon nanotubes with their helium-ion microscope.
- All of my current and former colleagues at the Institute of Nanotechnology, whom I have not already mentioned.
- All members of our get-together of the Physical Institute.

My family has been supporting me throughout my studies, helped whenever needed and always provided a safe haven. I am deeply grateful to them.

Birgit, I am so glad that we met during our time here. I hope we can build a future together.

Thank you!

Experimental Condensed Matter Physics (ISSN 2191-9925)

Herausgeber
Physikalisches Institut

Prof. Dr. Hilbert von Löhneysen
Prof. Dr. Alexey Ustinov
Prof. Dr. Georg Weiß
Prof. Dr. Wulf Wulfhekel

Die Bände sind unter www.ksp.kit.edu als PDF frei verfügbar oder als Druckausgabe bestellbar.

Band 1 Alexey Feofanov
Experiments on flux qubits with pi-shifters. 2011
ISBN 978-3-86644-644-1

Band 2 Stefan Schmaus
Spintronics with individual metal-organic molecules. 2011
ISBN 978-3-86644-649-6

Band 3 Marc Müller
Elektrischer Leitwert von magnetostriktiven Dy-Nanokontakten. 2011
ISBN 978-3-86644-726-4

Band 4 Torben Peichl
Einfluss mechanischer Deformation auf atomare Tunnelsysteme – untersucht mit Josephson Phasen-Qubits. 2012
ISBN 978-3-86644-837-7

Band 5 Dominik Stöffler
Herstellung dünner metallischer Brücken durch Elektromigration und Charakterisierung mit Rastersondentechniken. 2012
ISBN 978-3-86644-843-8

Band 6 Tihomir Tomanic
Untersuchung des elektronischen Oberflächenzustands von Ag-Inseln auf supraleitendem Niob (110). 2012
ISBN 978-3-86644-898-8

Band 7 Lukas Gerhard
Magnetoelectric coupling at metal surfaces. 2013
ISBN 978-3-7315-0063-6

Experimental Condensed Matter Physics
(ISSN 2191-9925)

Band 8 Kirill Fedorov
Fluxon readout for superconducting flux qubits. 2013
ISBN 978-3-7315-0067-4

Band 9 Jochen Zimmer
Cooper pair transport in arrays of Josephson junctions. 2014
ISBN 978-3-7315-0130-5

Band 10 Oliver Berg
Elektrischer Transport durch Nanokontakte von Selten-Erd-Metallen. 2014
ISBN 978-3-7315-0209-8

Band 11 Grigorij Jur'evic Grabovskij
Investigation of coherent microscopic defects inside the tunneling barrier of a Josephson junction. 2014
ISBN 978-3-7315-0210-4

Band 12 Cornelius Thiele
STM Characterization of Phenylene-Ethynylene Oligomers on Au(111) and their Integration into Carbon Nanotube Nanogaps. 2014
ISBN 978-3-7315-0235-7